THIOLS

STRUCTURE, PROPERTIES AND REACTIONS

CHEMISTRY RESEARCH
AND APPLICATIONS

Additional books and e-books in this series can be found
on Nova's website under the Series tab.

CHEMISTRY RESEARCH AND APPLICATIONS

THIOLS

STRUCTURE, PROPERTIES AND REACTIONS

CARLOS C. MCALPINE
EDITOR

nova
science publishers
New York

NOTICE TO THE READER

Library of Congress Cataloging-in-Publication Data

ISBN: 978-1-53615-599-0

Published by Nova Science Publishers, Inc. † New York

CONTENTS

PREFACE

The thiol (-SH) is an essential functional group in the biological system. The -SH groups play a catalytic and structural role in enzymes and proteins that participate in the defense against reactive oxygen species, detoxification, signal transduction, energy production, apoptosis, and other biological functions. In this compilation, the authors begin by exploring how the inorganic and organic forms of mercury interact with these macromolecules.

In the following study, thiol-methacrylate networks based on a tetrafunctional thiol and dimethacrylate monomers were prepared by both photopolymerization and amine-catalyzed Michael addition reaction. The progress of the polymerization reaction was monitored by FTIR and Raman spectroscopy. The cured materials were characterized by measuring the glass transition temperature, the flexural modulus, and the compressive strength.

In the closing study, a novel and efficient protocol is presented which has been developed for the synthesis of symmetrical disulfides from thiols using thionyl chloride as the sole oxidizing agent at ambient conditions. Also, a tentative mechanism has been reported for the reaction.

Chapter 1 - The thiol (-SH) is an essential functional group in the biological system. The -SH groups play a catalytic and structural role in enzymes and proteins that participate in the antioxidant system, detoxification, signal transduction, energy production, apoptosis, and other biological functions. The -SH group is present in several low molecular mass-SH (LMM-SH) molecules, such as homocysteine (HCys), cysteine (Cys), cysteinylglycine (CysGly), lipoic acid, coenzyme A (CoA), and glutathione (GSH). In the human plasma, the levels of high molecular mass-SH (HMM-SH) molecules range from 0.2–0.4 mM, while the LMM-SH molecules levels range from 0.1–20 μM. Since –SH groups are soft nucleophiles, they have a high affinity for soft electrophiles. In this context, both organic and inorganic forms of mercury (Hg) are soft electrophiles and, therefore, have a high affinity for -SH groups. The Hg is an environmental toxicant that has no biochemical or physiologic function in the organisms. The exposure to organic or inorganic Hg chemical forms is dangerous to the individuals, especially to the organisms in development. In this chapter, the authors will explore how the inorganic and organic forms of Hg interact with HMM-SH and HMM-SeH macromolecules.

Chapter 2 - Thiol-ene systems have attracted considerable recent interest because they display efficient, versatile and selective "click" reactions. In this study, thiol-methacrylate networks based on a tetrafunctional thiol (PTMP) and dimethacrylate monomers were prepared by both photopolymerization and amine-catalyzed Michael addition reaction. The progress of the polymerization reaction was monitored by FTIR and Raman spectroscopy. The cured materials were characterized by measuring the glass transition temperature, the flexural modulus and the compressive strength. Innovative hydrogels based on PTMP and a water-soluble dimethacrylate monomer (LBisEMA) were prepared by both visible light photo polymerization and Michael addition reaction. Depending on the synthesis method employed significant differences in the degree of swelling were observed. LBisEMA–PTMP hydrogels prepared by the Michael addition reaction catalyzed by propylamine resulted in the highest water uptake. Hydrogels prepared by visible light photo-polymerization contained unreacted thiol groups because of a faster homopolymerization reaction of the methacrylate groups. No significant effect of the PTMP/LBisEMA molar ratio on the water sorption capacity of the photocured hydrogel was observed. These trends are explained in terms of a balance between the mass fraction of hydrophilic groups and the crosslinking density of the network. Silver nanoparticles can be easily incorporated into thiol-ene systems because of the stabilizing properties of thiol functional groups. These polymer nanocomposites are very attractive for the preparation of biomaterials or dermatological patches with improved biocompatibility. Silver nanoparticles were prepared in PTMP by in situ reduction of silver nitrate with 2,6-di-tert-butyl-p-cresol. In this system, the thiol monomer acts as both stabilizing agent and reactive solvent. Mixtures of PTMP containing silver nanoparticles and a bifunctional methacrylate monomer were photoactivated with 2,2-dimethoxy-2-phenylacetophenone or camphor-quinone (CQ) and then photopolymerized by irradiation with UV or visible light respectively. Visible light photo polymerization is commonly carried out using the CQ-amine photoinitiator system. CQ displays an intense dark yellow color which bleaches under irradiation giving as a result colorless polymers. However, the authors found that in thiol-ene systems CQ is regenerated through hydrogen transfer reactions between ketyl radicals and thiyl radicals. This feature must be taking into account when considering the long-term color stability of the derived cured materials.

Chapter 3 - A novel and efficient protocol has been developed for the synthesis of symmetrical disulfides from thiols using thionyl chloride (SOCl$_2$) as a sole oxidizing agent at ambient conditions. Also, a tentative mechanism has been reported for the reaction. Compared to the already reported methods for the synthesis of disulfides, this method has several advantages, including short reaction time, no laborious work-up procedures (by-products are gases and escape from the reaction vessel promptly), a broad functional group tolerance, and takes place in solution as well as in solid states.

In: Thiols: Structure, Properties and Reactions   ISBN: 978-1-53615-599-0
Editor: Carlos C. McAlpine   © 2019 Nova Science Publishers, Inc.

Chapter 1

BIOLOGICAL THIOLS AND THEIR
INTERACTION WITH MERCURY

Cláudia S. Oliveira[1,], Pablo A. Nogara[2],*
Quelen I. Garlet[3], Guilherme S. Rieder[2]
and João B. T. Rocha[2,†]
[1]Instituto de Pesquisa Pelé Pequeno Príncipe, Curitiba, PR, Brazil
[2]Departamento de Bioquímica e Biologia Molecular, CCNE,
Universidade Federal de Santa Maria, Santa Maria, RS, Brazil
[3]Programa de Pós-Graduação em Farmacologia, CCS,
Universidade Federal de Santa Maria, Santa Maria, Brazil

ABSTRACT

The thiol (-SH) is an essential functional group in the biological system. The -SH groups play a catalytic and structural role in enzymes and proteins that participate in the antioxidant system, detoxification, signal transduction, energy production, apoptosis, and other biological functions. The -SH group is present in several low molecular mass-SH (LMM-SH)

* Corresponding Author's E-mail: jbtrocha@yahoo.com.br.
† Corresponding Author's E-mail: claudia.bioquimica@yahoo.com.br.

molecules, such as homocysteine (HCys), cysteine (Cys), cysteinylglycine (CysGly), lipoic acid, coenzyme A (CoA), and glutathione (GSH). In the human plasma, the levels of high molecular mass-SH (HMM-SH) molecules range from 0.2–0.4 mM, while the LMM-SH molecules levels range from 0.1–20 μM. Since –SH groups are soft nucleophiles, they have a high affinity for soft electrophiles. In this context, both organic and inorganic forms of mercury (Hg) are soft electrophiles and, therefore, have a high affinity for -SH groups. The Hg is an environmental toxicant that has no biochemical or physiologic function in the organisms. The exposure to organic or inorganic Hg chemical forms is dangerous to the individuals, especially to the organisms in development. In this chapter, we will explore how the inorganic and organic forms of Hg interact with HMM-SH and HMM-SeH macromolecules.

1. INTRODUCTION

The sulfur atom (S) is located in the group 16 (chalcogens) of Periodic Table, presenting oxidation states ranging from -2 to +6. The organic functional group thiol (-SH) is composed by the S atom bonded to an organic fragment (radical) and to a hydrogen atom (H). Thus, the –SH is analogous to the alcohol moiety (-OH) found in various biomolecules. Methionine (Met) and cysteine (Cys) are the S-containing amino acids found in the mammalian proteins. Cys, the –SH-containing amino acid, is analogous to the serine (Ser) and selenocysteine (Sec) amino acids, where the S atom is replaced by O and Se, respectively (Bhagavan and Ha 2015) (Figure 1). The –SH moiety is essential in the biological systems performing many functions, for instance, nucleophilic and redox catalysis, structural stabilization, allosteric regulation, posttranslational modification site, and metal coordination (Fomenko et al. 2008; Reddie and Carroll 2008; Zeida et al. 2014).

The -SH groups are also called mercaptans, a terminology coined by the Danish chemist William Christopher Zeise in 1934. The term "mercaptan" is derived from the Latin *corpus mercurium captans*, which mean "a body that captures mercury" due to the high affinity of S for mercury (Hg) (Jensen 1989; Goldwater 1965). In ancient times, the alchemists described the strong affinity of S for Hg and pondered the misture of both as the start material for the creation of new metals (Newman 2014).

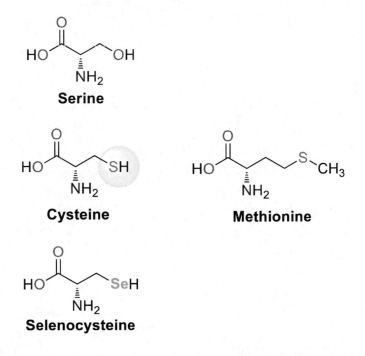

Figure 1. The structural formula of Met, Cys, Ser and Sec. The -SH group of Cys is highlighted.

The Hg toxicity caught attention of the general public after the Minamata (Harada 1995; Ekino et al. 2007) and Iraq outbreaks (Amin-Zaki et al. 1976; Smith et al. 1976). The organic Hg poisoning is a critical health hazard since it can permanently harm the brain physiology, (Jackson 2018; Oliveira et al. 2018a; Nogara et al. 2019). Previous biochemical and toxicological studies have identified the high molecular mass-SH or -SeH (HMM-SH/SeH) molecules as the principal targets of Hg. The interactions of Hg species with these macromoleculesmay cause loss of function, which is followed by cellular injury (Bjørklund et al. 2017; Oliveira et al. 2017). The Hg neurotoxic effects can be severe depending on the level of exposure and period of contamination. For example, when the exposure occurs during the fetal development, the damage is irreversible due to the immaturity of the central nervous system (CNS) and blood-brain barrier (BBB) (Clarkson et al. 2003; Ren et al. 2017). Here, in this chapter, we will explore the Hg interactions with endogenous low molecular mass -SH (LMM-SH) and HMM-SH molecules, as well as the effects of these interactions in the organisms.

2. THE CHEMISTRY OF –SH

The -SH group plays a vital role in biological systems due to its reactivity toward oxidant agents and electrophiles (including metals, such as Cu, Fe, Zn, Cd, Pb, and Hg) (Reddie and Carrol 2008; Poole 2015). The reactivity of -SH can be attributed to the electron rich S atom and to its large size (d-orbitals allow for multiple oxidation states), low dissociation energy of the S-H bond, and low oxidation-reduction (redox) potential (Table 1) (Reddie and Carrol 2008; Zeida et al. 2014). The equilibrium thiol/thiolate (-SH $\leftrightarrow$ -S$^-$ + H$^+$) is essential since the negatively charged -S$^-$ moiety is a better nucleophile than the neutral -SH moiety (Reddie and Carrol 2008; LoPachin and Gavin 2014). Thus, the reactivity of Cys depends on the pKa value of its –SH (Table 1) and on the medium where the protein is localized (Zeida et al. 2014; Poole 2015). The Cys pKa values can range from 3.5 to >10.5, depending on the protein considered.

Table 1. Comparison of the amino acids properties

Amino acid	pKa	Bond dissociation energy (kcal/mol)[c]	Covalent radii (pm)[c]	Redox potential (V)
Serine	~13.6[a]	428 (O-H)	66 (O)	Not Found
Cysteine	~8.5[b]	344 (S-H)	106 (S)	-0.27 to -0.12[d]
Selenocysteine	~5.3[b]	305 (Se-H)	117 (Se)	-0.31to -0.27[e]

[a] (Bruice et al. 1961); [b] (Byun and Kang 2011; Poole 2015) [c] (Dean 1999); [d] (Reddie and Carrol 2008; Li et al. 2014); [e] (Müller et al. 1994; Besse et al. 1997; Metanis et al. 2006).

These shift in pKa values can result from the interactions of the -SH moiety with other groups (protein backbone, amino acid residues,and solvent molecules) and the dielectric properties of the medium (Harris and Turner 2002; Li et al. 2005).

According to the Pearson's theory (Hard and Soft Acids and Bases - HSAB) the -SH moiety is considered a soft base/nucleophile and consequently exhibit affinity by soft acids/electrophiles (Pearson 1963, 1990; LoPachin et al. 2012). A Lewis base is characterized by the presence of an available electron lone pair, the capacity of electron donation can indicate the nucleophilicity of the chemical species. The nucleophilicity increases according to the increase in the group of the Periodic Table, due to the increase in the orbital polarizability. Thus, the nucleophilicity order: Ser < Cys < Sec, is associated with the increase of the atomic radius size from O to Se (Table 1; Figure 2). Also, the atomic radius size and orbital polarization can be used to characterize if a chemical species will behave as a hard or soft nucleophile. In this way, the Ser can behave as a hard base while the Cys and Sec behave as soft bases (Bunnett 1957, 1963; Ibne-Rasa 1967; Galembeck and Caramori 2003; Bachrach et al. 2004; Bento and Bickelhaupt 2008).

The Cys modification (oxidation, adduction formation, and reduction) in biological systems are necessary to maintainvital physiological intracellular pathways, for instance, antioxidant responses, redox signaling, posttranslational modifications, and gene regulation (Walsh et al. 2005; Reddie and Carrol 2008; Chalker et al. 2009; Bischoff and Schlüter 2012; Couvertier et al. 2014; Poole 2015). The products of these modifications are disulfide bonds (R-SS-R), sulfenic (R-SOH), sulfinic (R-SO$_2$H), sulfonic (R-SO$_3$H) acids derivatives, S-nitrosothiols (R-SNO) and S-glutathionylation (R-SSG) (Figure 3). The modified proteins R-SOH and R-SNO can react with the reduced glutathione (GSH) forming the stable S-glutathionylated species (R-SSG). The R-SSG and R-SNO proteins can be reduced back to the native thiol form (R-SH) by the glutaredoxin/ glutaredoxin reductase (Grx/GrxR) and thioredoxin/thioredoxin reductase (Trx/TrxR) systems, using NADPH as cofactor (Forman et al. 2010; Antelmann and Helmann 2011; Klomsiri et al. 2011; Chung et al. 2013; Couvertier et al. 2014; Poole 2015; Gusarov and Nudler 2018). In this scenario, the interaction of these redox systems with exogenous electrophiles (i.e., Hg chemical forms) can disrupt the cellular function leading to the cellular apoptosis and metabolism failure (Ihara et al. 2017; Oliveira et al. 2018a).

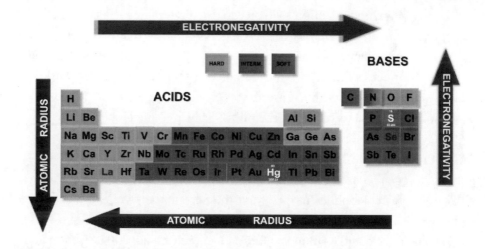

Figure 2. Hard and Soft acids and bases following Pearson's theory.

3. BIOLOGICAL –SH

Intracellularly the concentration of LMM-SH ranges from 2 – 5 mM whereas, in the plasma, the concentration ranges from 230 to 340 µM (Wardman et al. 1992; Levy et al. 1993). The chief representatives of LMM-SH molecules found in the extra- and intracellular medium are Cys, GSH, homocysteine (HCys), γ-L-glutamyl-L-cysteine (γ-Glu–Cys), and L-cysteinyl-L-glycine (Cys-Gly) (Table 2; Figure 4 and 5). These molecules are found mainly in the reduced form (-SH) in the cellular milieu while in the plasma the majority of them are in the oxidized form (Turell et al. 2013).

The thiol/disulfide (-SH/-SS-) networks comprised mainly Cys, GSH, glutaredoxin (Grx), glutathione reductase (GR), Trx and TrxR are kept reduced mainly with reducing equivalents derived from NADPH (Figure 6). The –SH redox balance is critical for multiple metabolic, signaling and transcriptional processes in mammalian cells (Figure 6). The HMM-SH molecules have several functions, for instance, the transport of ions and inorganic/organic molecules, and regulation of cell process (growth and proliferation) and cell redox state (Table 2). The albumin (Alb) is the most abundant –SH-containing molecule in plasma (~0.6 mM) and ~75% of this molecule is in the reduced form (Turell et al. 2013). Also, hydrogen sulfide (H_2S), the smallest –SH-containing molecule, is present in the blood in a low concentration (~15 nM). This physiological messenger has reactivity similar to those of the LMM-SH molecules (Kabil and Banerjee 2010).

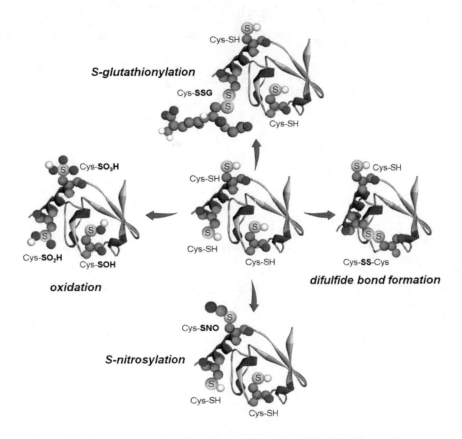

Figure 3. Most important covalent cysteinyl residues modifications in proteins.

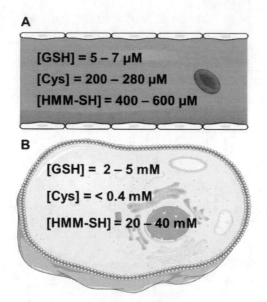

Figure 4. Plasma (A) and intracellular (B) thiol levels. Data according to Wardman et al. 1992, Levy et al. 1993, Turell et al. 2013, Prakash et al. 2009.

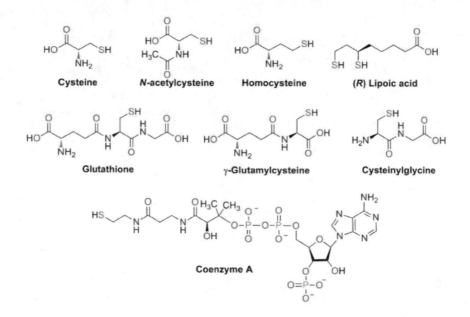

Figure 5. Main LMM-SH found in mammalian organisms.

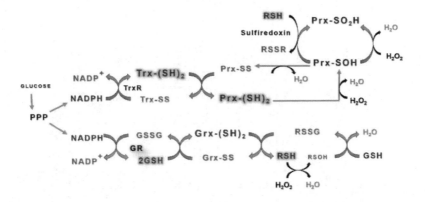

Figure 6. Cellular redox couples in the biological systems. The highlighted molecules are potential targets of Hg chemical forms. -SH and -SS- means the reduced and oxidized form of the molecule, respectively. PPP = Pentose phosphate pathway; NADP$^+$ = Nicotinamide adenine dinucleotide phosphate oxidized; NADPH = Nicotinamide adenine dinucleotide phosphate reduced; Trx = Thioredoxin; TrxR = Thioredoxin reductase; Prx = Peroxiredoxin; GSH = Glutathione; GR = Glutathione reductase; Grx = Glutaredoxin; R = any L/HMM-SH molecule.

Figure 7 and Table 2 represent some of the critical HMM-SH molecules, which can be a target for oxidant agents and metals. The oxidation of HMM-SH can cause loss of function and consequently, cell injury. As mentioned, the presence of reduced -SH groups could facilitate the binding of toxic metals (such as Hg and Cd) in these proteins.

3.1. Cys Motifs

The Cys residues can be found in several proteins forming different motifs, such as: XCX, XCCX, XCCCX, CXC, CCXCC, CXCC, CCXC, CXCXC, CXCXCXC, CXXC, CXXXC and CXXXXC (X can be any amino acid residue) (Alexandroff et al. 1998; Woycechowsky and Raines 2003; Wouters et al. 2010; Knerlich-Lukoschus and Held-Feindt 2015; Ziller et al. 2017; Oliveira et al. 2018a). The CC, CXC and CXXC motifs are often present in thiol oxidoreductases proteins, for example, thioredoxins, glutaredoxins, protein disulfide isomerases, chemokines, among others (Wouters et al. 2007, 2010; Fomenko et al. 2008; Bulleid and Ellgaard 2011).

Table 2. Main -SH-containing molecules (LMM-SH and HMM-SH) found in the mammalian organisms

LMM-SH				
Molecule	**Function**	**[ ]**	**Mass (Da)**	**Ref.**
Cys	Redox balance in cells; Neuroprotection; Precursor of GSH, hydrogen sulfide, and taurine.	Plasma 202–281 μM	121	Mansoor et al. 1992; Andersson et al. 1993; Kleinman and Richie 2000; Giustarini et al. 2005, 2015; Nolin et al. 2007; McBean 2017; Paul et al. 2018
GSH	Scavenger of reactive oxygen and nitrogen species; Detoxification of xenobiotics; Regulation, modulation, and maintenance of cellular redox homeostasis.	Plasma 4.9–7.3 μM	307	Mansoor et al. 1992; Andersson et al. 1993; Kleinman and Richie 2000; Giustarini et al. 2005, 2015; Nolin et al. 2007; Couto et al. 2016; McBean 2017; Paul et al. 2018; Farina and Aschner 2019
γ-Glu–Cys	Antioxidant; Anti-inflammatory; Intermediate of GSH synthesis; Modulates the expression of proteins related to antioxidant defense.	Plasma 3.1–5.4 μM	250	Andersson et al. 1993; Kleinman and Richie 2000; Nakamura et al. 2012; Ferguson and Bridge 2016; Yang et al. 2019
Coenzyme A	Synthesis and oxidization of fatty acids; Oxidation of pyruvate in the citric acid cycle; Metabolic intermediate and second messenger; Control of cellular processes: energy metabolism, mitosis, and autophagy.	Plasma 9×10^{-6} μM Cell 0.467 nmol/mg prot	767	Leonardi et al. 2005; Choudhary et al. 2014; Pietrocola et al. 2015; Shurubor et al. 2017
Homocysteine (HCys)	Modulation of growth of cells and tissues, protein structural modifications and oxidative stress; It may exert toxic effects on brain; Acts as an agonist of glutamate receptors and NMDA receptors, which leads to an increase in cytoplasmic calcium, high production of free radicals, and activation of pro-apoptotic caspases; Independent risk factor for thrombosis.	Plasma 5.0-11.9 μM Urine 1.0-2.2 mmol/mol creatinine	135	Mansoor et al. 1992; Andersson et al. 1993; Mattson and Shea 2003; Alexander et al. 2013; Lippi et al. 2014

LMM-SH

Molecule	Function	[]	Mass (Da)	Ref.
Lipoic acid	Antioxidant; Detoxification reactive oxygen species; Metal scavenger; Cofactor for mitochondrial enzymes; Catalysation of the oxidative decarboxylation of pyruvate, α-ketoglutarate and branched-chain α-keto acids; Modulation of the NF-κB; Attenuates the release of cytotoxic cytokines.	Plasma 1–25 ng/mL	206	Biewenga et al. 1997; Kataoka 1998; Navari-Izzo et al. 2002; Booker 2004; Gorąca et al. 2011; Cronan, 2018
Oxidized glutathione (GSSG)	GSH oxidized form; Substrate for GSH production through reduction; Indicator of cell functionality and oxidative stress.	Plasma 1.4-3.2 μM	612	Monostori et al. 2009; Turell et al. 2013; Giustarini et al. 2015
N-acetylcysteine	Precursor of GSH; Antioxidant; Mucolytic agent; Treatment of paracetamol intoxication and toxicity-induced by pesticides.	Blood 4.29-5.0 μM	163	Longo et al. 1991; Elbini Dhouib et al. 2016
Cysteinylglyc ine	Derived from GSH; Biomarker in autism; Associated with phenylketonuria.	Plasma 18.6-35.8 μM Urine 1.7-1.9 mmol/mol creatinine	178	Perry and Hansen 1981; Kuśmierek et al. 2009; McMenamin et al. 2009; Frustaci et al. 2012

HMM-SH

Molecule	Function	[]	Mass (kDa)	Ref.
Trx	Catalysation disulfide/dithiol changes; Transference of electrons to Prx, methionine sulfoxide reductase, and redox-sensitive transcription factors; Acts as intracellular reductase, promoting cellular proliferation; Apoptosis inhibition through direct binding to apoptosis signal-regulating kinase-1.	Plasma 10 -80 ng/mL/	11.7	Masutani et al. 2005; Turell et al. 2013; Lu and Holmgren 2014; Branco and Carvalho 2018

Table 2. (Continued)

Molecule	Function	HMM-SH [ ]	Mass (kDa)	Ref.
Alb	Main protein of plasma; Binding and transport of Ca^{2+}, Na^+, K^+, Zn^{2+}, fatty acids, hormones, bilirubin, heavy metals (Cd^{2+}, Co^{2+}, Ni^{2+}, Hg^{2+}) and xenobiotics; Maintenance of acid-base balance; Regulation of blood colloidal osmotic pressure.	Blood 600 μM	65–70	Lu et al. 2008; Infusino and Panteghini 2013; Bern et al. 2015; Kumar and Banerjee 2017
MT	Comprised of MT-I and MT-II, MT-III and MT-IV classes; Binding capacity available for Se, Cd^{2+}, Hg^{2+}, Cu^+, Cu^{2+}, Ag^+, Au^+, Bi^{+3}, As^{+3}, Co^{2+}, Fe^{+2}, Pb^{2+}, Pt^{2+}, Tc^{4+} and xenobiotics; Antioxidants; Involved in metalloregulatory processes that include cell growth and multiplication; Detoxification of the heavy metals; Up-regulation linked to carcinogen processes.	Liver 100 μg/g Kidney 933 μg/g	0.5–14	Onosaka et al. 1985, 1986; Szczurek et al. 2001; Thirumoorthy et al. 2007; Capdevila et al. 2012
δ-AlaD	Involved in the heme biosynthesis. Main lead-binding protein in erythrocytes. Urinary ALA (U-ALA) has been used as a biomarker for lead exposure.	295 U	36	Warren et al. 1998; Bergdahl et al. 1997; Beyer et al. 2013; Tasmin et al. 2015; Kerr et al. 2019
Hk	Involved in ATP and hormones binding; Promote glucose phosphorylation.	Muscle 4.33 U/g	102	Ritov and Kelley 2001; Calmettes et al. 2015
Hb	Major component of mammalian erythrocytes; Oxygen-transporter; Mediates NO signaling (NO interact with the cysteine at the beta-93 position to form S-nitroshemoglobin, which inhibits NO bioactivity).	Blood 120–180 g/dL	646	Beutler and Waalen 2006; Helms and Shapiro 2013; Farzam et al. 2018; Diepstraten and Hart 2019
PDI	Thiol oxidoreductase chaperone; Catalyses the formation of disulfide bonds; Plays a role in thrombosis, contributing to coagulation activation.	Endoplasmic reticulum 0.2–0.5 mM	54	Lyles and Gilbert 1991; Watanabe et al. 2014; Soares Moretti and Martins Laurindo 2017; Chen et al. 2018

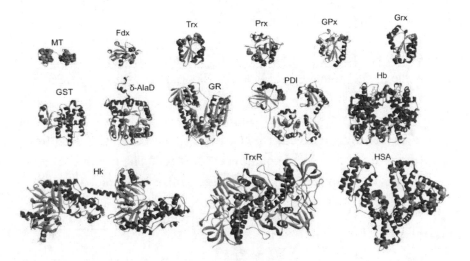

Figure 7. Selected HMM-SH/-SeH molecules found in organisms. The Cys and Sec residues are highlight in CPK model. Metallothionein - MT (PDB 4MT2); Ferredoxin - Fdx (PDB 3P1M); Thioredoxin - Trx (PDB 1ERT); Peroxiredoxin - Prx (PDB 1OC3); Glutathione Peroxidase - GPx (PDB 1GP1); Glutaredoxin - Grx (PDB 2FLS); Glutathione S-Transferase - GST (PDB 19GS); δ-Aminolevulinic acid dehydratase - δ-AlaD (PDB 5HMS); Glutathione Reductase - GR (PDB 1GRA); Protein Disulfide Isomerases - PDI (PDB 4EKZ); Hemogloblin - Hb (PDB 5KSI); Hexokinase – Hk (PDB 2NZT); Thioredoxin Reductase - TrxR (PDB 2J3N); Human serum albumin - HSA (PDB 1H9Z). The number of Cys residues in the proteins can change according to the isoforms or organisms considered. References are presented in the Table 3.

The MT, a well-conserved Cys-rich protein, contains a significant number of different Cys motifs (Braun et al. 1992; Ziller et al. 2017) (Table 3). The XCX and CXXC motifs are the most abundant found in the human and mouse proteome (Oliveira et al. 2018a) and CXXC usually is present in the N-terminus of α-helical structures (Iqbalsyah et al. 2006; Conway and Lee 2015). The Cys residues reactivity in these motifs depends on their position inside the protein backbone and the chemical environment, which will influence the pKa values of the thiol groups. In general, the most reactive Cys is found closest to the N-terminus of a given protein, presenting a pKa value ranging from 3.5 – 7.5, and consequently, the Cys residue is in the deprotonated form (-S⁻) (Weichsel et al. 1996; Mavridou et al. 2014; Conway and Lee 2015; Poole 2015; Netto et al. 2007). The XCX is found in the caspases (cysteine proteases), which have an essential role in inflammation and programmed cell death (apoptosis, pyroptosis, and necroptosis). In the caspase active site, the Cys residue performs a specific nucleophilic attack on the peptide bond of specific aspartate residues cleaving the target protein (Nicholson and Thornberry 1997; Clark 2016).

Table 3. Number of Cys, Sec, and motifs present in some of HMM-SH that are potential targets of Hg chemical species. The Cys residues from the redox center were considered reduced. The number of Cys residues in parenthesis indicates the number of reduced thiol

Protein	N° Cys	N° Sec	N° motif	PDB ID	Ref.
δ-AlaD	9 (9)	0	1 CXC; 1 CXXC; 6 XCX	5HMS	Mills-Davies et al. 2017
Alb	35 (1)	0	8 CC; 19 XCX	1H9Z	Petitpas et al. 2001
Fdx	5 (5)	0	2 CXXC; 1 XCX	3P1M	Chaikuad 2010*
Grx	4 (2)	0	1 CXXC; 2 XCX	2FLS	Johansson 2006*
GPx	4 (4)	1	4 XCX; 1 XUX	1GP1	Epp et al. 1983
GR	9 (9)	0	1 CXXXXC; 7 XCX	1GRA	Karplus and Schulz 1989
GST	4 (4)	0	4 XCX	19GS	Oakley et al. 1997
Hb[a]	6 (6)	0	6 XCX	5KSI	Sun et al. 2017
Hk	25 (25)	0	25 XCX	2NZT	Rabeh 2006*
MT	20 (20)	0	3 CC; 7 CXC; 3 CXXC; 4 CXXXC	4MT2	Braun et al. 1992
Prx	3 (3)	0	3 XCX	1OC3	Evrard 2004
PDI	6 (6)	0	2 CXXC; 2 XCX	4EKZ	Wang et al. 2013
Trx	5 (5)	0	1 XCX; 1 CXXC; 1 CXXXC	1ERT	Weichsel et al. 1996
TrxR[b]	26 (26)	2	2 CU; 20 XCX; 2 CXXXXC	2J3N	Fritz-Wolf et al. 2007

[a]tetramer; [b]dimer; *paper not published.

Moreover, the Cys domains C_2H_2 (Cys_2His_2), C_3H (Cys_3His), and C_4 (Cys_4), known as zinc fingers, have diverse functions in proteins, such as DNA recognition, transcriptional activation, regulation of apoptosis, protein folding and assembly (Laity et al. 2001; Krishna et al. 2003; Maret et al. 2006; Eom et al. 2012). The term "zinc finger" was created to describe a nine-fold repeated Cys- and His-rich sequence motif of the eukaryotic transcription factor IIIA (Pelham and Brown 1980; Klug and Schwabe 1995).

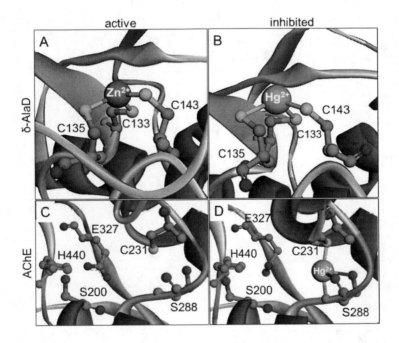

Figure 8. Comparison between the δ-AlaD enzyme complexed with Zn^{2+} (A) and Hg^{2+} (B), and the native AChE (C) and Hg-inhibited AChE (D). Despite the similar binding mode and molecular geometry, the δ-AlaD + Hg^{2+} do not present catalytic active (Erskine et al. 2000; Rocha et al. 2012). The binding of Hg^{2+} in the fish AChE active site (represented by the catalytic triad: S200, H440, and E327) cause its inhibition (distance Ser-O···Hg = 8 Å). The proteins were obtained from Protein Data Bank - PDB (http://www.rcsb.org/) (δ-AlaD + Zn^{2+}: PDB ID 1H7O (Erskine et al. 2001); δ-AlaD + Hg^{2+}: PDB ID 1QML (Erskine et al. 2000); AChE: PDB ID 1EA5 (Dvir et al. 2002); AChE + Hg^{2+}: PDB ID 2J4F (Kreimer et al. 1994). The figures were obtained from Discovery Studio Visualizer software -DSV (http://www.3dsbiovia.com/).

In this way, all of these Cys motifs and domains can be targets for oxidant species, for example, the electrophilic Hg forms (E^+Hg) (Hartwig 2001; Aschner et al. 2006; Witkiewicz-Kucharczyk and Bal 2006; Carvalho AT et al. 2008; Kröncke and Klotz 2009; Sutherland and Stillman 2011; Xu et al. 2013; Oliveira et al. 2018a; Nogara et al. 2019) (Table 3). Besides the fact that Hg chemical forms could bind to the -SH/-SeH groups of proteins, such as, TrxR (Wagner et al. 2010; Rodrigues et al. 2015), acetylcholinesterase (AChE) (Kreimer et al. 1994) creatine kinase (CK) (Glaser et al. 2010), and hexokinase (Hk) (Kanda et al. 1976), they can replace the physiological metal cofactor, as observed for δ-AlaD (Figure 8) (Erskine et al. 2000; Rocha et al. 2012) and MT (Lu et al. 1993; Leiva-Presa et al. 2004). In fact, in some proteins, the Hg is found bound to $-S^-$, showing tetra-, tri-, bi-, and monocoordinated complexes (Holm et al. 1996; Nogara et al. 2019).

3.2. Biological -SH Oxidation

As discussed above the oxidation -SH groups can cause critical changes in the cell milieu, either inactivating or activating metabolic pathways (Dhanbhoora and Babson 1992; Aoshiba et al. 1999; Haouzi et al. 2001; Limón-Pacheco et al. 2007).

The reactive oxygen species (ROS), such as hydrogen peroxide (H_2O_2), peroxynitrite (ONOOH), superoxide anion ($O_2^{\bullet-}$) and hydroxyl radical ($HO^\bullet$), are considered the primary players of oxidative stress in the living cells. The oxidative damage can involve the oxidation of proteins, lipids, and nucleic acid and can be modulated by antioxidants species (Netto et al. 2007; Birben et al. 2012; Vazquez-Torres 2012; Sueishi et al. 2014; Thanan et al. 2015; Davies 2016). In this sense, the -SH-containing molecules play an essential role in cell protection against oxidative injury (Sies 1991; Winterbourn and Metodiewa 1999; Winterbourn and Hampton 2008; Pisoschi and Pop 2015; Short et al. 2016). Table 4 indicates the $HO^\bullet$ as the most reactive ROS, while the H_2O_2 reacts slowly with -SH groups (e.g., Cys, GSH and Alb) (reactive order: $HO^\bullet \gg O_2^{\bullet-} > ONOOH \gg H_2O_2$).

According to the formation constant Cys and GSH with different metals (Table 5), it is possible to see that the different Hg chemical forms present high affinity by LMM-SH, which can lead to -SH oxidation and loss of function as will be discussed in the next sections of this chapter.

Table 4. Reaction rate constant, k ($M^{-1} s^{-1}$), of –SH-containing molecules with oxidant species

Oxidant	L/HMM-SH	k ($M^{-1} s^{-1}$)	Ref.
H_2O_2	Cys	$2.9 - 4.7$	Radi et al. 1991; Winterbourn and Metodieva 1999
	GSH	0.87	Winterbourn and Metodieva 1999
	Alb	$1.4 - 2.26$	Carballal et al. 2003; Torres et al. 2012
ONOOH	Cys	4.5×10^3	Radi et al. 1991; Quijano et al. 1997
	GSH	1.36×10^3	Trujillo and Radi 2002
	Alb	$7.5 - 9.7 \times 10^3$	Alvarez et al. 1999; Torres et al. 2012
$O_2^{\bullet-}$	Cys	2.7×10^6	Asada and Kanematsu 1976
	GSH	$1.1 \times 10^3 - 6.7 \times 10^5$	Asada and Kanematsu 1976; Sueishi et al. 2014
	Alb	*	
$HO^{\bullet}$	Cys	$2.1 - 5.35 \times 10^9$	Masuda et al. 1973; Mezyk 1996a
	GSH	$3.48 \times 10^9 - 1.64 \times 10^{10}$	Mezyk 1996b; Sueishi et al. 2014
	Alb	8×10^{10}	Davies 2016

Table 5. Formation constant (log K) for -SH molecules and metals

Metal	Specie	Cys	GSH
Zn^{2+}	ML	9.4 [a]	7.6 [a]
	ML_2	18.6 [a]	12.9 [a]
Cd^{2+}	ML	10.9 [a]	8.3 [a]
	ML_2	17.0 [a]	12.9 [a]
Hg^{2+}	ML	14.2 [b]	26.0 [f]
	ML_2	38.6 [c]	37.5 [c]
Pb^{2+}	ML	12.2 [d]	10.5 [d]
	ML_2	18.6 [d]	15.0 [d]
$MeHg^+$	ML	11.6 [e]	11.6 [e]

M = Metal, and L = thiol ligand; [a]Walsh and Ahner 2013; [b]Lenz and Martell 1964; [c]Stricks and Kolthoff 1953; [d]Corrie et al. 1976; [e]Reid and Rabenstein 1981; [f]Oram et al. 1996. The formation constant can vary according to the authors' methodology.

4. THIOL AFFINITY FOR HG

4.1. Mercury

Mercury (Hg) is a transition metal located in the group 12 of the Periodic Table (Tinkov et al. 2015). The Hg is widely distributed in the environment, and the exposure to any chemical form of Hg is expected to have toxicological consequences (Oliveira et al. 2017, 2018a; Nogara et al. 2019). On the other hand, due to Hg unique chemical characteristics and versatility, this metal has been employed in the manufacture of medical devices (Horowitz et al. 2014), fluorescent lamps (Li et al. 2018), among others (Ho et al. 2017; Li et al. 2019). The primary concern about the Hg usage is its uncontrolled discard in the wastewater or even the Hg-containing devices, which summed with the artisanal and small-scale gold mining (ASGM) have been contributing to the increase of the Hg environmental levels (Muntean et al. 2014, 2018). In this context, Streets et al. (2018) demonstrated that the United States, Canada, and Europe have been decreasing the anthropogenic emission of Hg since 2010. Conversely, Asia, Africa, Central America, and Oceania increased the Hg emission in the same period. Thus, the aftermath of the global amount of Hg released by anthropogenic sources increased by 1.7% per year since 2010, and the main contributors for this increase were cement production and ASGM (Streets et al. 2018).

Another important aspect about the environmental toxicity of Hg that has emerged in the last two decades is its long persistence in the environment (Veiga and Hinton 2002; Maramba et al. 2006; Cooke et al. 2009, 2013; Amos et al. 2013; Streets et al. 2017; Brown et al. 2018; Hsu-Kim et al. 2018; Rudd et al. 2018). Indeed, the intense debate about the legacy of Hg in different ecosystems (Liu et al. 2019; Morris et al. 2019; Olsen et al. 2019; Xu et al. 2019; Zhang et al. 2019) clearly indicates that Hg toxicity will endure even after the abandonment of its use by the man.

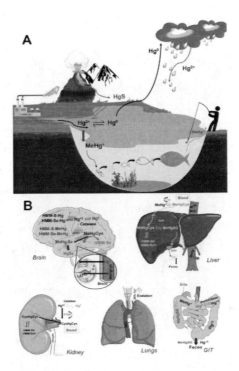

Figure 9. Fate of E$^+$Hg in the environment (A) and in the mammalian primary target organs (B). Although different Hg chemical forms are naturally found in the environment, the anthropogenic activities have potentialized the environmental levels of Hg in the last decades. Once metabolized to organic mercury, MeHg$^+$, it bioaccumulates in the aquatic web chain and can reach the human population that lives close to the contaminated regions. Into the body, the E$^+$Hg toxicokinetic and toxicodynamics is dictated by the interactions and exchange reactions with -SH-containing molecules. The Hg^0 has as first target the lungs, and through the catalase enzyme, it is oxidized to Hg^{2+} that rapidly bound to –SH or -SeH. The bind of Hg^{2+} with two LMM-SH molecules (mainly Cys) drives the complex LMM-S-Hg-S-LMM (for example, Cys-Hg-Cys) to the kidneys. Although the MeHg$^+$ accumulates mainly in the liver, the major toxicological concern about MeHg$^+$ is it capacity to cross the blood-brain barrier and reach the CNS. The MeHg$^+$ enters into the brain as a mimic of the amino acid Met through the LAT transporter and by others not yet well-studied mechanisms. Note that though E$^+$Hg species (Hg^{2+} and MeHg$^+$) are represented as the free ionic forms, their existence of both in the biotic and abiotic system are more likely to be associated either with -SH (organic thiol groups) or with sulphide (S^{-2}, HS$^-$, H$_2$S or "inorganic thiol group") or selenide (Se^{-2}, HSe$^-$, H$_2$Se or "inorganic selenol group") (for review see Nogara et al. 2019).

The elemental mercury (Hg^0) can easily be oxidized to Hg^{2+} in the environment by the microorganism present in the sediment or by O$_3$ and Br in the atmosphere (Beckers and Rinklebe 2017). Shreds of evidence demonstrated that the Hg^0 oxidation into the body is mediated by the enzyme catalase (Magos et al. 1978; Eide and Syversen 1982; Ogata et al. 1987). Additionally, once Hg^{2+} is formed it is expected to bind rapidly to -SH or -SeH (Nogara et al. 2019). The Hg^{2+}, formed in the environment or released by anthropogenic actions can be methylated by some microorganism, especially bacteria that express the gene cluster hgcAB (Gilmour et al. 2013; Parks et al. 2013). The $MeHg^+$ formation, bioaccumulation, and biomagnification in the aquatic web chain are of great concern, since predatory fish (which can accumulate significant amounts of $MeHg^+$) may be part of the human diet (Figure 9).

The methylation of Hg^{2+} in the mammalian body might occur in the gastrointestinal tract by the microbiota; however, the amount of Hg methylated in the gastrointestinal tract is negligible when compared with the amount of Hg that is methylated by the aquatic microorganism and accumulated in fish (Gilmour et al. 2013; Martín-Doimeadios et al. 2017). Nevertheless, the demethylation of $MeHg^+$ in the mammalian body has been in the focus of the toxicological researches. Some studies described the accumulation of inorganic mercury in mammalian organs after the exposure to $MeHg^+$ (Vahter et al. 1994; Charleston et al. 1995; Burbacher et al. 2005; Ishitobi et al. 2010), and more recently, studies demonstrated that the dominant chemical form of inorganic Hg accumulated is an insoluble complex with Se (HgSe, mercury selenide) (Wagemann et al. 2000; Kehrig et al. 2008; Huggins et al. 2009; Sakamoto et al. 2015). Theoretical and practical studies demonstrated that the selenoamino acids can facilitate the breaking of the C-Hg bond (Khan and Wang 2010; Asaduzzaman and Schreckenbach 2011); however, the mechanism of $MeHg^+$ demethylation and HgSe formation in the mammalian body is still unclear.

4.2. Chemistry of Hg and –SH/-S⁻

In conformity with Pearson's theory, the electrophilic Hg species (E^+Hg), i.e., Hg^{2+}, $MeHg^+$, and $EtHg^+$, are soft acids, and -SH/-S⁻ species are soft bases (Figure 2). These species present a high affinity for each other, with the stability constant up to 10^{40} (Pearson 1963, 1990; Casas and Jones 1980; Dyrssen and Wedborg 1991; Melnick et al. 2010; Merle et al. 2018; Mesterházy 2018). At the atomic level, the 4p atomic orbital of S and the 6s orbital of Hg are involved in the S-Hg bond (for the Cys-S-Hg^{2+} complex) (Belcastro et al. 2005).

Despite the high stability of the S-Hg bond, it is labile, i.e., in the presence of other -SH (or -SeH) groups, ligand exchange reactions can easily occur (Reaction 1 and 2; Figure 10) (Fuhr and Rabenstein 1973; Rabenstein 1978; Rabenstein and Evans 1978, Rabenstein and Reid 1984; Nogara et al. 2019). Nogara et al. (2019) recently named these type of exchange reactions as the Rabenstein's Reaction in honor to Dr. Dallas Rabenstein, the pioneer in the study of E$^+$Hg exchange reactions.

$$R^1\text{-S-Hg-S-}R^2 + R^3\text{-SH} \rightarrow R^3\text{-S-Hg-S-}R^2 + R^1\text{-SH}$$

Reaction 1. Exchange reaction of -SH with Hg^{2+} complex (R = Cys, GSH, protein).

$$R^1\text{-S-HgMe} + R^2\text{-SH} \rightarrow R^2\text{-S-HgMe} + R^1\text{-SH}$$

Reaction 2. Exchange reaction of -SH with MeHg$^+$ complex (R = Cys, GSH, protein).

The high affinity of E$^+$Hg for -SH/-S$^-$ drives the behavior of Hg inside the body. Hence, the bind of Hg with the biological -SH moiety dictates the fate of Hg in the body, as well as its toxicity (Oliveira et al. 2017, 2018; Nogara et al. 2019). On the other hand, molecules rich in -SH have been used on the process of Hg detoxification in both the environment and mammalian body (Basinger et al. 1981; Rooney 2004; Bridges et al. 2009; Baran 2010; Li et al. 2011; He et al. 2012). As corollary, one may suggest that Rabenstein's reactions dictate the Hg absorption, distribution, and excretion, i.e., once inside the organism, the Hg^{2+} and/or MeHg$^+$ can suffer several exchange reactions until it finds a specific target (for example, HMM-SH or HMM-SeH) (Rabenstein 1978; Nogara et al. 2019).

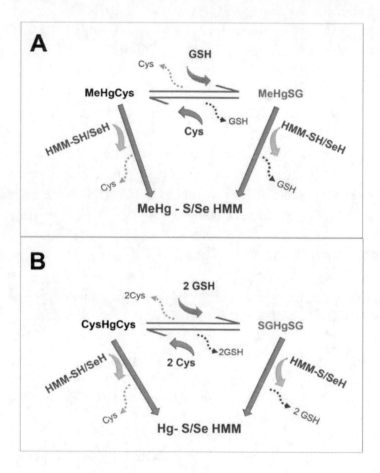

Figure 10. Practical example of the Rabenstein's reactions into the body.

4.3. The Toxicokinetic and Toxicodynamics of E⁺Hg into the Body Depends on -SH Molecules

Based on the formation constant of electrophilic forms of Hg (E⁺Hg) with –SH groups, free E⁺Hg species are not expected to be found inside the body. Except for the stomach, where the low pH can favor the formation of

Hg salt with Cl⁻, i.e., $HgCl_2$ and MeHgCl (Rabentein 1978; Nogara et al. 2019), in the other organs, the E⁺Hg species are predominantly bound to LMM-SH (Cys or GSH) or to HMM-SH/-SeH. The bind of E⁺Hg species to -SH-containing molecules will influence the toxicokinetics and toxicodynamics of Hg in a given cell or organism. Most important, the toxicokinetics and toxicodynamics of Hg vary depending on the Hg chemical form.

After undergoing Rabenstein's reaction in the bloodstream, Hg species are transported bound to Alb, selenoprotein P or LMM-SH (Cys and GSH) (Bridges and Zalups 2005; Liu et al. 2018). Due to the lability of E⁺Hg and -SeH/-SH interactions, several exchange reactions may happen before the E⁺Hg forms reach and remain in a given target organ. It is well established that E⁺Hg species cross through the plasma membrane as a mimic of some amino acids, namely, Met and cystine (Bridges and Zalups 2005). As depicted in Figure 11, the bind of Hg^{2+} to two cysteines, molecules forms a molecule structurally similar with the amino acid cysteine, while the bind of MeHg⁺ with one cysteine is analogous to the amino acid Met. These amino acids have specific transporters located in the cell membrane that facilitate the entrance into the cell milieu. The organic anion transporter (OAT) and L-type amino acid transporter (LAT) are the transporters of cystine and MeT and likely the Cys-Hg-Cys and MeHg-Cys (Bridges and Zalups 2005). Studies demonstrate that high concentration of cystine or Met in the extracellular medium decrease the Hg^{2+} and MeHg⁺ uptake by the cells (Zalups and Lash 2006; Roos et al. 2011; Zimmermann et al. 2013), proving the role of the amino acid transporters in the Hg influx.. The multidrug resistance protein 2 (MRP2) facilitate the E⁺Hg species excretion in the urine or the feces. The E⁺Hg forms cross through the MRP2 transporter bound to one or two LMM-SH (Cys or GSH) molecules or the synthetic Hg-chelating agents, DMSA and DMPS (Aleo et al. 2005; Bridges et al. 2008a,b, 2011; Zalups and Bridges 2009; Oliveira et al. 2018b).

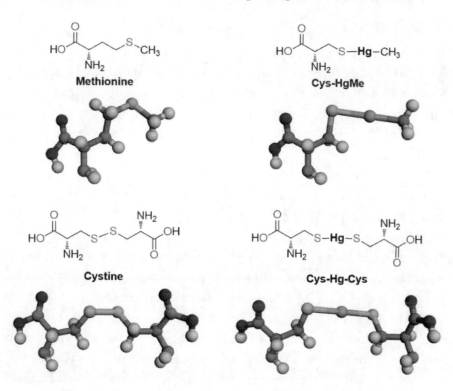

Figure 11. Structural comparison between endogenous thiol-molecules (Methionine and Cystine) and its Hg mimics (Cys-HgMe and Cys-Hg-Cys). The 3D molecules were built from Avogadro software (https://avogadro.cc/).

Once in the bloodstream, the toxicodynamic of E^+Hg forms will vary depending on the chemical form. The inorganic ion form, Hg^{2+}, accumulates mainly in the kidney. Although the organic form, $MeHg^+$, accumulates in the liver and kidneys, the major toxicological concerns are associated with its accumulation in the brain tissue, where it can induce irreversible damage (Oliveira et al. 2017, 2018a; Nogara et al. 2019). The Table 6 shows the different distribution of Hg^{2+} and $MeHg^{2+}$ in Wistar rats exposed intravenously to E^+Hg, as well as demonstrates the effects of Hg interaction to -SH-containing molecules on the Hg body distribution. Generally, the exposure to Hg complexed with LMM-SH (Cys and GSH) increases the Hg renal uptake.

Table 6. Dose, time and Hg chemical form deposition in blood, liver, and kidney after intravenous injection

Mercury (dose)	Time after exposure	Organ (% of the dose)			Animal (weight)	Ref.
HgCl$_2$ (0.25 µmol/kg)		Blood	Liver	Kidney	Male Sprague Dawley rats (200g)	Zalups and Barfuss 1993, 1995a,b
	2 min	33%	5%	5%		
	5 min	35%	6%	5%		
	20 min	25%	7%	11%		
	1 h	19%	7%	15%		
	3 h	9%	10%	39%		
	24 h	2%	9%	45%		
HgCl$_2$ (0.5 µmol/kg)		Blood	Liver	Kidney	Male Wistar rats (200g)	Zalups 1993; Bridges et al. 2014; Oliveira et al. 2018b
	1 min	42%	5%	5%		
	5 min	30%	5%	10%	Male Sprague Dawley rats (200g)	
	15 min	42%	6%	10%		
	30 min	18%	8%	24%		
	1 h	22%	5%	45%		
	3h	8%	6%	55%		
	24 h	2%	3%	57%		
	48h	1%	5%	40%		
HgCl$_2$ (1.8 µmol/kg)		Blood	Liver	Kidney	Male Wistar rats (180g)	Zalups and Barfuss 1996
	1h	14%	13%	32%		
	24h	1%	6%	43%		
GS-Hg-SG (0.25 µmol/kg)		Blood	Liver	Kidney	Male Sprague Dawley rats (200g)	Zalups and Barfuss 1995a
	5 min	17%	6%	7%		
	1 h	10%	8%	22%		
	24 h	1%	8%	50%		
Alb-Hg-Alb (0.25 µmol/kg)		Blood	Liver	Kidney	Male Sprague Dawley rats (200g)	Zalups and Barfuss 1993
	2 min	29%	6%	6%		
	5 min	23%	5%	9%		
	20 min	15%	6%	12%		
	3 h	6%	9%	36%		
Cys-Hg-Cys (0.5 µmol/kg)	48h		Liver 2%	Kidney 59%	Male Wistar rats (200g)	Bridges et al. 2008b
Hcy-Hg-Hcy (0.5 µmol/kg)	48h		Liver 3%	Kidney 58%	Male Wistar rats (200g)	Bridges et al. 2008b
Cys-Hg-Cys (1.8 µmol/kg)		Blood	Liver	Kidney	Male Wistar rats (180g)	Zalups and Barfuss 1996
	1h	8%	7%	44%		
	24h	1%	6%	52%		
MeHgCl (5 mg/kg)	48h	Blood 41%	Liver 8%	Kidney 5%	Male Wistar rats (200g)	Bridges et al. 2011
MeHg-Cys (5 mg/kg)	48h	Blood 25%	Liver 6%	Kidney 18%	Male Wistar rats (200g)	Bridges et al. 2011

Table 7. IC$_{50}$ of E$^+$Hg with potential molecular targets

Enzyme	Hg specie	IC$_{50}$ (µM)	Organ/ tissue	Specie	Ref.
AChE	Hg^{2+}	13	Brain	*Parachromis managuensis*	Araújo et al. 2016
AChE	Hg^{2+}	220	Brain	*Cichla ocellaris*	Silva et al. 2012
AChE	Hg^{2+}	13	Brain	*Hoplosternum littorale*	Araújo et al. 2018
Ac. phosph.	Hg^{2+}	139	Liver	*Metynnis. argenteus*	Dantzger et al. 2017
Alk. phosph.	Hg^{2+}	10^{-11}	Cells	*Arthrospira platensis*	Tekaya et al. 2012
ADC	MeHg$^+$ EtHg$^+$ PhHg$^+$ Hg^{2+}	0.00796 0.070 0.145 0.364	PC12 cells	*Rattus norvegicus*	Wang et al. 2016
NAT1	MeHg$^+$ Hg^{2+}	20 3	Lung ephitelial cells	*Homo Sapiens*	Ragunathan et al. 2010
Ca^{2+}-ATPase	Hg^{2+}	0.75 0.85	Heart Gills	*Saccobranchus fossilis*	Bansal et al. 1984
H$^+$-ATPase	Hg^{2+}	0.60	Purified cell membrane	*Nitellopsis obtusa*	Manusadzianas et al. 2002
Na$^+$, K$^+$-ATPase	Hg^{2+}	**0.50**	Gill	*Micropterus salmoides*	Jagoe et al. 1996
δ-ALAD	Hg^{2+}	~120 ~60 ~70	Liver Kidney Brain	*Wistar rats*	Peixoto et al. 2004
FPG	MeHg$^+$ Hg^{2+}	0.0043 4.04	Cells	*Escherichia coli*	Wu et al. 2015
β-galactosidase	Hg^{2+}	0.76	Purified enzyme	*Aspergillus oryzae*	Zhang et al. 2018
GPx	MeHg$^+$	1.9	Erythorcytes	*Bovinus*	Farina et al. 2009
GR	Hg^{2+}	509	Liver	*Oncorhynchus mykiss*	Tekman et al. 2013
CACT	MeHg$^+$ Hg^{2+}	0.137 0.09	Liver	*Danio rerio*	Tonazzi et al. 2018
ORC	MeHg$^+$ Hg^{2+}	0.38 0.21	Purified enzyme	*Homo Sapiens*	Giangregorio et al. 2018
Paraoxonase	Hg^{2+}	490	Blood	*Scyliorhinus canicula*	Sayın et al. 2012
PKC	MeHg$^+$ Hg^{2+}	0.22 1.5	Brain	Sprague-Dawley rats	Rajanna et al. 1995
TrxR	MeHg$^+$ Hg^{2+}	0.0197 $7.2×10^{-3}$	Purified enzyme	*Homo sapiens*	Carvalho CML et al. 2008

AchE: Acetylcholinesterase; Ac. phosph: Acid phosphatase; Alk phosph. Alkaline phosphatase; ADC: Arginine decarboxylase; NAT1: Arylamine N-acetyltransferase 1; FPG: Formamidopyrimidine DNA glycosylase; GR: Glutathione reductase; CACT: Mitochondrial Carnitine/Acylcarnitine Transporter; ORC: Ornithine/citrulline; PKC: Protein kinase C; TrxR: Thioredoxin reductase.

5. THE E⁺HG HMM-SH TARGETS

As described above, Hg has a high affinity for -SH, and as a corollary, the main targets of E⁺Hg are –SH-containing molecules. The bind of Hg with -SH dictate the transport (Section 4) and toxicity of E⁺Hg. The Hg interaction with HMM-SH molecules can trigger a cascade of events that start with the binding of E⁺Hg species to HMM-SH, followed by conformational changes in the HMM-S-Hg⁺E complex and loss of function of the targed protein. The final consequences of such type of interaction of E⁺Hg and HMM-SH molecules can include a myriad of detrimental biological effects, for instance, an increase in the oxidative stress status, followed by cell death and organ failure.

Several studies have been done to search for enzymes inhibited by E⁺Hg species, in order to find the Hg preferential target(s) (Table 7). The E⁺Hg IC_{50} (the half maximal inhibitory concentration) ranging from 10^{-11} to 509 μM (Hg^{2+}) and 7.96×10^{-3} to 20 μM ($MeHg^+$), indicates considerable variability in the E⁺Hg biological effects. In this sense, we could suspect that the enzymes with high IC_{50} shown little affinity by Hg, while enzymes with low IC_{50} are potential targets for this metal. For example, the TrxR showed IC_{50} values of 7.2 and 19.7 nM for Hg^{2+} and $MeHg^+$, respectively (Table 7), which suggest that this molecule is a Hg target. The TrxR tridimensional structure contains 26 Cys and 2 Sec residues, besides 2 CU, 20 XCX and 2 CXXXXC motifs (Table 3), showing several possible Hg binding sites. Moreover, the GPx presented a high IC_{50} (1.9 μM) and a low number of motifs (4 XCX and 1 XUX) when compared with TrxR (Table 3 and 7). However, other factors can be involved in the enzyme inhibition by E⁺Hg, such as the accessibility of them in the Cys-binding regions (motifs and domains), i.e., due to the complex tertiary/quaternary protein structures, not all the Cys residues (and/or Sec) are available to react with the Hg species. In this way, the steric hindrance has a vital role in the selectivity of protein interaction to Hg. Another aspect worth to be mentioned is the source of – SH-containing enzymes (e.g., crude extract *versus* purified proteins). The IC_{50} values for a given enzyme in crude extracts usually are much higher than for the same purified enzyme.

In addition, the -SH rich molecules also, are used as a marker of Hg exposure, for instance, several studies have been reported the enzyme δ-ALA-D as a marker of divalent metals (including Hg) exposure in experimental and occupational contamination cases (Hernberg et al. 1971; Schütz and Skerfving 1975; Grotto et al. 2010; Franciscato et al. 2011; Rocha et al. 2012).

Some studies have indicated the enzyme AChE as a marker of fish Hg exposure (Silva et al. 2013; Araújo et al. 2016, 2018). However, the toxic effects of mercurials differ from an organism to another. When a free Cys residue is present in the AChE, as in *Torpedo californica*, the Hg^{2+} inhibition is irreversible and occurs in the µM range. However, when the free Cys is not sensitive to Hg^{2+} (*Drosophila melanogaster*) or is otherwise absent (*Electrophorus electricus*), the inhibition occurs in the mM range (Frasco 2007). The difference in the AChE inhibition by Hg^{2+} in the *Torpedo californica* could be due to the interaction with the Cys residue 231 (C231; Figure 5). On the other hand, this residue is replaced to Gly in the human AChE (G234; Cheung et al. 2012), which could indicate that the human enzyme is less sensitive to Hg. The replacement of some amino acids in the protein sequence for a specific enzyme from different organisms is often found in the protein structures evolution, and it is involved in the specificity of its inhibitors (Lipman et al. 2002; Siltberg-Liberles et al. 2011; Hatton and Warr 2015).

CONCLUSION

The toxicological effects of E^+Hg are related to its interactions with –SH/SeH--containing molecules, where the metal is absorbed, distributed and excreted mainly by -SH exchange reactions, also named the Rabenstein's reactions. Despite the highest affinity of Hg to Se, the interactions of this metal with L/HMM-SH molecules are more relevant due to the higher concentration of these molecules compared to HMM-SeH molecules. In this way, there are several cellular targets for E^+Hg, which make the study of the molecular mechanism of E^+Hg toxicity extremely complex. Regardless of the *in vivo* and *in vitro* studies, the identification of E$^+$Hg primary toxic targets has still to be clarified. The use of *in silico* methods combined with omics approaches seems to be an essential strategy in the analysis and prediction of the E$^+$Hg species toxicity mechanisms. Also, the development of advanced analytical methods is necessary to improve the biochemical assays related to the identification of targets and elucidation of the Hg mechanism of action.

ACKNOWLEDGMENTS

This work was supported by Coordination for Improvement of Higher Education Personnel CAPES/PROEX (n°23038.005848/2018-31; n°0737/ 2018; n°88882.182123/2018-01), the National Council for Scientific and Technological Development (CNPq), as well as the Rio Grande do Sul Rio Grande do Sul Foundation for Research Support (FAPERGS - Brazil).

REFERENCES

Aleo, M. F., Morandini, F., Bettoni, F., Giuliani, R., Rovetta, F., Steimberg, N., Apostoli, P., Parrinello, G. and Mazzoleni, G. (2005) Endogenous thiols and MRP transporters contribute to Hg^{2+} efflux in HgCl$_2$-treated tubular MDCK cells. *Toxicology*, 206: 137-151.
Alexander, B., Ekaterina, B., Anna, M. and Elizaveta, V. (2013) Why is homocysteine toxic for the nervous and immune systems? *Current Aging Science*, 6: 29-36.
Alexandroff, A. B., McIntyre, C. A., Porter, J. C., Zeuthen, J., Vile, R. G. and Taub, D. D. (1998) Sticky and smelly issues: lessons on tumour cell and leucocyte trafficking, gene and immunotherapy of cancer. *British Journal of Cancer*, 77: 1806-1811.

Alvarez, B., Ferrer-Sueta, G., Freeman, B. A. and Radi, R. (1999) Kinetics of peroxy nitrite reaction with amino acids and human serum albumin. *The Journal of Biological Chemistry*, 274: 842-848.

Amin-Zaki, L., Elhassani, S., Majeed, M. A., Clarkson, T. W., Doherty, R. A., Greenwood, M. R. and Giovanoli-Jakubezak, T. (1976) Perinatal methylmercury poisoning in Iraq. *The American Journal of Diseases of Children*, 130: 1070-1076.

Amos, H. M., Jacob, D. J., Streets, D. G. and Sunderland, E. M. (2013) Legacy impacts of all-time anthropogenic emissions on the global mercury cycle. *Global Biogeochemical Cycles*, 27: 410-421.

Andersson, A., Isaksson, A., Brattstrom, L. and Hultberg, B. (1993). Homocysteine and other thiols determined in plasma by HPLC and thiol-specific post column derivatization. *Clinical Chemistry*, 39: 1590-1597.

Antelmann, H. and Helmann, J. D. (2011) Thiol-based redox switches and genes regulation. *Antioxidants and Redox Signaling*, 14: 1049-1063.

Aoshiba, K., Nakajima, Y., Yasui, S., Tamaoki, J. and Nagai, A. (1999) Red blood cells inhibit apoptosis of human neutrophils. *Phagocytes*, 93: 4006-4010.

Araújo, M. C., Assisa, C. R. D., Silva, K. C. C., Souza, K. S., Azevedo, R. S., Alves, M. H. M. E., Silva, L. C., Silva, V. L., Adame, M. L., Junior, L. B. C., Bezerra, R. S. and Oliveira M. B. M. (2018) Characterization of brain acetylcholinesterase of bentonic fish Hoplosternum littorale: Perspectives of application in pesticides and metal ions biomonitoring. *Aquatic Toxicology*, 205: 213-226.

Araújo, M. C., Assisa, C. R. D., Silva, L. C., Machado, D. C., Campos. K. C. S., Araújo, A. V. L., Carvalho Jr, L. B., Bezerra, R. S. and Oliveira M. B. M. (2016) Brain acetylcholinesterase of jaguar cichlid (*Parachromis managuensis*): From physicochemical and kinetic properties to its potential as biomarker of pesticides and metal ions. *Aquatic Toxicology*, 177: 182-189.

Asada, K. and Kanematsu, S. (1976) Reactivity of thiols with superoxide radicals. *Agricultural and Biological Chemistry*, 40: 1891-1892.

Asaduzzaman, A. M. and Schreckenbach, G. (2011) Degradation mechanism of methylmercury selenoamino acid complexes: A computational study. *Inorganic Chemistry*, 50: 2366-2372.

Aschner, M., Syversen, T., Souza, D. O. and Rocha, J. B. (2006) Metallothioneins: mercury species-specific induction and their potential role in attenuating neurotoxicity. *Experimental Biology and Medicine*, 231: 1468-1473.

Bachrach, S. M., Demoin, D. W., Luk, M. and Miller Jr, J. V. (2004) Nucleophilic attack at selenium in diselenides and selenosulfides. A computational study. *The Journal of Physical Chemistry A*, 108: 4040-4046.

Bansal, S. K., Murthy, R. C., and Chandra S. V. (1985) The effects of some divalent metals on cardiac and branchial Ca^{2+}-ATPase in a freshwater fish *Saccobranchus fossilis*. *Ecotoxicology and Environmental Safety*, 9: 373-377.

Baran, E. J. (2010) Chelation therapies: A chemical and biochemical perspective. *Current Medicinal Chemistry*, 17: 3658-3672.

Basinger, M. A., Casas, J. S., Jones, M. M., Weaver, A. D. and Weinstein, N. H. (1981) Structural requiriments for Hg(II) antidotes. *Journal of Inorganic and Nuclear Chemistry*, 43: 1419-1425.

Beckers, F. and Rinklebe, J. (2017) Cycling of mercury in the environment: sources, fate, and human health implications: A review. *Critical Reviews in Environmental Science and Technology*, 47: 693-794.

Belcastro, M., Marino, T., Russo, N. and Toscano, M. (2005) Interaction of cysteine with Cu^{2+} and Group IIb (Zn^{2+}, Cd^{2+}, Hg^{2+}) metal cations: a theoretical study. *Journal of Mass Spectrometry*, 40: 300-306.

Bento, A. P. and Bickelhaupt F. M. (2008) Nucleophilicity and leaving-group ability in frontside and backside SN2 reactions. *The Journal of Organic Chemistry*, 73: 7290-7299.

Bergdahl, I. A., Grubb, A., Schütz, A., Desnick, R. J., Wetmur, J. G., Sassa, S. and Skerfving, S. (1997) Lead binding to δ-aminolevulinic acid dehydratase (ALAD) in human erythrocytes. *Pharmacology & Toxicology*, 81: 153-158.

Bern, M., Sand, K. M. K., Nilsen, J., Sandlie, I. and Andersen, J. T. (2015) The role of albumin receptors in regulation of albumin homeostasis: Implications for drug delivery. *Journal of Controlled Release*, 211: 144-162.

Besse, D., Siedler, F., Diercks, T., Kessler, H. and Moroder, L. (1997) The redox potential of selenocystine in unconstrained cyclic peptides. *Angewandte Chemie*, 36: 883-885.

Beutler, E. and Waalen, J. (2006) The definition of anemia: what is the lower limit of normal of the blood hemoglobin concentration? *Blood* 107: 1747-1750.

Beyer, W. N., Franson, J. C., French, J. B., May, T., Rattner, B. A., Shearn-Bochsler, V. I., Warner, S. E., Weber, J. and Mosby, D. (2013) Toxic exposure of songbirds to lead in the Southeast Missouri Lead Mining District. *Archives of Environmental Contamination and Toxicology*, 65: 598-610.

Bhagavan, N. V. and Ha, C. (2015) *Amino acids*. Chapter 3, p. 21-29.

Biewenga, G. P., Haenen, G. R. M. M. and Bast, A. (1997) The pharmacology of the antioxidant lipoic acid. *General Pharmacology*, 29: 315-331.

Birben, E., Sahiner, U. M., Sackesen, C., Erzurum, S. and Kalayci, O. (2012) Oxidative stress and antioxidant defense. *World Allergy Organization Journal*, 5: 9-19.

Bischoff, R. and Schlüter, H. (2012) Amino acids: Chemistry, functionality and selected non-enzymatic post-translational modification. *Journal of Proteomics*, 75: 2275-2296.

Bjørklund, G., Aaseth, J., Ajsuvakova, O. P., Nikonorov, A. A., Skalny, A. V., Skalnaya, M. G. and Tinkov, A. A. (2017). Molecular interaction between mercury and selenium in neurotoxicity. *Coordination Chemistry Reviews*, 332: 30-37.

Booker, S. J. (2004) Unraveling the pathway of lipoic acid biosynthesis. *Chemistry and Biology*, 11: 10-12.

Branco, V. and Carvalho, C. (2018) The thioredoxin system as a target for mercury compounds. *Biochimica et Biophysica Acta - General Subjects*, doi: 10.1016/j.bbagen.2018.11.007.

Braun, W., Vasák, M., Robbins, A. H., Stout, C. D., Wagner, G., Kägi, J. H. and Wüthrich, K. (1992) Comparison of the NMR solution structure and the x-ray crystal structure of rat metallothionein-2. *Proceedings of the National Academy of Sciences of the United States of America*, 89: 10124-10128.

Bridges, C. C. and Zalups, R. K. (2005) Molecular and ionic mimicry and the transport of toxic metals. *Toxicology and Applied Pharmacology*, 204: 274-308.

Bridges, C. C., Joshee, L. and Zalups, R. K. (2008a) Multidrug resistance proteins and the renal elimination of inorganic mercury mediated by 2,3-dimercaptopropane-1-sulfonic acid and meso-2,3-dimer-captosuccinic acid. *Journal of Pharmacology and Experimental Therapeutics*, 324: 383-390.

Bridges, C. C., Joshee, L. and Zalups, R. K. (2008b) MRP2 and the DMPS- and DMSA-mediated elimination of mercury in TR⁻ and control rats exposed to thiol S-conjugates of inorganic mercury. *Toxicological Sciences*, 105: 211-220.

Bridges, C. C., Joshee, L. and Zalups, R. K. (2014) Aging and the deposition and toxicity of mercury in rats. *Experimental Gerontology*, 53: 31-39.

Bridges, C., C., Joshee, L. and Zalups, R. K. (2009) Effect of DMPS and DMSA on the placental and fetal disposition of methylmercury. *Placenta*, 30: 800-805.

Bridges, C., C., Joshee, L. and Zalups, R. K. (2011) MRP2 and the handling of mercuric ions in rats exposed acutely to inorganic and organic species of mercury. *Toxicology and Applied Pharmacology*, 251: 50-58.

Brown, T. M., Macdonald, R. W., Muir, D. C. and Letcher, R. J. (2018) The distribution and trends of persistent organic pollutants and mercury in marine mammals from Canada's Eastern Arctic. *Science of the Total Environment*, 618: 500-517.

Bruice, T. C., Fife, T. H., Bruno, J. J. and Brandon, N. E. (1962) Hydroxyl group catalysis. II. The reactivity of the hydroxyl group of serine. The nucleophilicity of alcohols and the ease of hydrolysis of their acetyl esters as related to their pKa'. *Biochemistry*, 1: 7-12.

Bulleid, N. J. and Ellgaard, L. (2011) Multiple ways to make disulfides. *Trends in Biochemical Sciences*, 36: 485-492.

Bunnett, A. (1957) New factor affecting reactivity in bimolecular nucleophilic displacement reactions. *Journal of the American Chemical Society*, 79: 5969-5974.

Bunnett, A. (1963) Nucleophilic reactivity. *Annual Review of Physical Chemistry*, 14: 271-290.

Burbacher, T. M., Shen, D. D., Liberato, N., Grant, K. S., Cernichiari, E. and Clarkson, T. (2005) Comparison of blood and brain mercury levels in infant monkeys exposed to methylmercury or vaccines containing thimerosal. *Environmental Health Perspectives*, 113: 1015-1021.

Byun, B. J. and Kang, Y. K. (2011) Conformational preferences and pK(a) value of selenocysteine residue. *Biopolymers*, 95: 345-353.

Calmettes, G., Ribalet, B., John, S., Korge, P., Ping, P. and Weiss, J. N. (2015) Hexokinases and cardioprotection. *Journal of Molecular and Cellular Cardiology*, 78: 107-115.

Capdevila, M., Bofill, R., Palacios, Ò. and Atrian, S. (2012) State-of-the-art of metallothioneins at the beginning of the 21st century. *Coordination Chemistry Reviews*, 256: 46-62.

Carballal, S., Radi, R., Kirk, M. C., Barnes, S., Freeman, B. A. and Alvarez, B. (2003) Sulfenic acid formation in human serum albumin by hydrogen peroxide and peroxynitrite. *Biochemistry*, 42: 9906-9914.

Carvalho, A. T., Swart, M., van Stralen, J. N., Fernandes, P. A., Ramos, M. J. and Bickelhaupt, F. M. (2008) Mechanism of thioredoxin-catalyzed disulfide reduction. Activation of the buried thiol and role of the variable active-site residues. *The Journal of Physical Chemistry B*, 112: 2511-2523.

Carvalho, C. M. L., Chew, E., Hashemy, S. I., Lu, J. and Holmgren, A. (2008) Inhibition of human thioredoxin system: A molecular mechanism of mercury toxicity. *The Journal of Biological Chemistry*, 283: 11913-11923.

Casas, J. S. and Jones, M. M. (1980) Mercury (II) complexes with sulfhydryl containing chelating agents: stability constant inconsistencies and their resolution. *Journal of Inorganic and Nuclear Chemistry*, 42: 99-102.

Chaikuad, A., Johansson, C., Krojer, T., Yue, W.W., Phillips, C., Bray, J.E. et al. (2010) *Crystal structure of human ferredoxin-1 (FDX1) in complex with iron-sulfur cluster.* doi: 10.2210/pdb3P1M/pdb.

Chalker, J. M., Bernardes, G. J., Lin, Y. A. and Davis, B. G. (2009) Chemical modification of proteins at cysteine: opportunities in chemistry and biology. *Chemistry, an Asian Journal,* 4: 630-640.

Charleston, J. S., Body, R. L., Mottet, N. K., Vahter, M. E. and Burbacher, T. M. (1995) Auto metallographic determination of inorganic mercury distribution in the cortex of the calcarine sulcus of the monkey *Macaca fascicularis* following long-term subclinical exposure to methylmercury and mercuric chloride. *Toxicology and Applied Pharmacology,* 132: 325-333.

Chen, F., Zhao, Z., Zhou, J., Lu, Y., Essex, D. W. and Wu, Y. (2018) Protein disulfide isomerase enhances tissue factor-dependent thrombin generation. *Biochemical and Biophysical Research Communications,* 501: 172-177.

Cheung, J., Rudolph, M. J., Burshteyn, F., Cassidy, M. S., Gary, E. N., Love, J., Franklin, M. C. and Height, J. J. (2012) Structures of human acetylcholinesterase in complex with pharmacologically important ligands. *Journal of Medicinal Chemistry,* 55: 10282-10286.

Choudhary, C., Weinert, B. T., Nishida, Y, Verdin, E. and Mann, M. (2014) The growing landscape of lysine acetylation links metabolism and cell signalling. *Nature Reviews,* 15: 536-550.

Chung, H. S., Wang, S. B., Venkatraman, V., Murray, C. I. and Van Eyk. J. E. (2013) Cysteine oxidative posttranslational modifications: emerging regulation in the cardiovascular system. *Circulation Research,* 112: 382-392.

Clark, A. C. (2016) Caspase allostery and conformational selection. *Chemical Reviews,* 116: 6666-6706.

Clarkson, T. W., Magos, L. and Myers, G. J. (2003) Human exposure to mercury: the three modern dilemmas. *The Journal of Trace Elements in Experimental Medicine,* 16: 321-343.

Conway, M. E. and Lee, C. (2015) The redox switch that regulates molecular chaperones. *Biomolecular Concepts,* 6: 269-284.

Cooke, C. A., Balcom, P. H., Biester, H. and Wolfe, A. P. (2009) Over three millennia of mercury pollution in the Peruvian Andes. *Proceedings of the National Academy of Sciences*, 106: 8830-8834.

Cooke, C. A., Hintelmann, H., Ague, J. J., Burger, R., Biester, H., Sachs, J. P. and Engstrom, D. R. (2013) Use and legacy of mercury in the Andes. *Environmental Science & Technology*, 47: 4181-4188.

Corrie, A. M., Walker, M. D. and Williams, D. R. (1976) Thermodynamic considerations in co-ordination. Part XXII. Sequestering ligands for improving the treatment of plumbism and cadmiumism. *Journal of the Chemical Society, Dalton Transactions*, 0: 1012-1015.

Couto, N., Wood, J. and Barber, J. (2016) The role of glutathione reductase and related enzymes on cellular redox homoeostasis network. *Free Radical Biology and Medicine*, 95: 27-42.

Couvertier, S. M., Zhou, Y. and Weerapana, E. (2014) Chemical-proteomic strategies to investigate posttranslational modifications. *Biochimica et Biophysica Acta*, 1844: 2315-2330.

Cronan, J. E. (2018) Advances in synthesis of biotin and assembly of lipoic acid. *Current Opinion in Chemical Biology*, 47: 60-66.

Dantzger, D. D., Dantzger, M., Jonsson, C. M. and Aoyama, H. (2017) In vitro effects of agriculture pollutants on microcrustacean and fish acid phosphatases. *Water Air, and Soil Pollution*, doi: 10.1007/s11270-017-3570-7.

Davies, M. J. (2016) Protein oxidation and peroxidation. *Biochemical Journal*, 473: 805-825.

Dean, J. A. (1999) *Large's handbook of chemistry*. Fifteenth Edition. McGRAW-HILL, INC.

Dhanbhoora, C. M. and Babson, J. R. (1992) Thiol depletion induces lethal cell injury in cultured cardiomyocytes. *Archives of Biochemistry and Biophysics*, 293: 130-139.

Diepstraten, S. T. and Hart, A. H. (2019). Modelling human haemoglobin switching. *Blood Reviews*, 33: 11-23.

Dvir, H, Wong, D. M., Harel, M., Barril, X., Orozco, M., Luque, F. J., Muñoz-Torrero, D., Camps, P., Rosenberry, T. L., Silman, I. and Sussman, J. L. (2002) 3D structure of *Torpedo californica* acetylcholinesterase complexed with huprine X at 2.1 A resolution: kinetic and molecular dynamic correlates. *Biochemistry*, 41: 2970-2981.

Dyrssen, D. and Wedborg, M. (1991) The sulphur-mercury(II) system in natural waters. *Water, Air, and Soil Pollution*, 56: 507-519.

Eide, I. and Syversen, T. L. M. (1982) Uptake of elemental mercury and activity of catalase in rat, hamster, guinea-pig, normal and acatalasemic mice. *Acta Pharmacologica and Toxicologica*, 51: 371-376.

Ekino, S., Susa, M., Ninomiya, T., Imamura, K. and Kitamura, T. (2007). Minamata disease revisited: an update on the acute and chronic manifestations of methyl mercury poisoning. *Journal of the Neurological Sciences* 262: 131-144.

Elbini Dhouib, I., Jallouli, M., Annabi, A., Gharbi, N., Elfazaa, S. and Lasram, M. M. (2016). A mini review on N-acetylcysteine: An old drug with new approaches. *Life Sciences*, 151: 359-363.

Eom, K. S., Cheong, J. S. and Lee, S. J. (2016) Structural analyses of zinc finger domains for specific interactions with DNA. *Journal of Microbiology and Biotechnology*, 26: 2019-2029.

Epp, O., Ladenstein, R. and Wendel, A. (1983) The refined structure of the selenoenzyme glutathione peroxidase at 0.2-nm resolution. *European Journal of Biochemistry*, 133: 51-69.

Erskine, P. T., Duke, E. M., Tickle, I. J., Senior, N. M., Warren, M. J. and Cooper, J. B. (2000) MAD analyses of yeast 5-aminolaevulinate dehydratase: their use in structure determination and in defining the metal-binding sites. *Acta Crystallographica D*, 56: 421-430.

Erskine, P. T., Newbold, R., Brindley, A. A., Wood, S. P., Shoolingin-Jordan, P. M., Warren, M. J. and Cooper, J. B. (2001) The x-ray structure of yeast 5-aminolaevulinic acid dehydratase complexed with substrate and three inhibitors. *Journal of Molecular Biology*, 312: 133-141.

Evrard, C., Capron, A., Marchand, C., Clippe, A., Wattiez, R., Soumillion, P., Knoops, B. and Declercq, J.-P. (2004) Crystal structure of a dimeric oxidized form of human peroxiredoxin 5. *Journal of Molecular Biology*, 337: 1079-1090.

Farina, M. and Aschner, M. (2019) Glutathione antioxidant system and methylmercury-induced neurotoxicity: An intriguing interplay.

Biochimica et Biophysica Acta General Subjects, doi: 10.1016/j.bbagen.2019.01.007.

Farina, M., Campos, F., Vendrell, I., Berenguer, J., Barzi, M., Pons, S. and Suñol, C. (2009) Probucol increases glutathione peroxidase-1 activity and displays long-lasting protection against methylmercury toxicity in cerebellar granule cells. *Toxicology Sciences*, 112: 416-426.

Farzam, P., Tamborini, D., Zimmermann, B., Wu, K. C., Boas, D. A. and Franceschini, M. A. (2018) Novel diffuse correlation spectroscopy for simulations estimation of hemoglobin concentration, oxygen saturation, and blood flow. *Biophotonics Congress: Biomedical Optics Congress 2018. Optical Society of America*, Hollywood, Florida, p. JTh3A.33.

Ferguson, G. and Bridge, W. (2016) Glutamate cysteine ligase and the age-related decline in cellular glutathione: The therapeutic potential of γ-glutamylcysteine. *Archives of Biochemistry and Biophysics*, 593: 12-23.

Fomenko, D. E., Marino, S. M. and Gladyshev, V. N. (2008) Functional diversity of cysteine residues in proteins and unique features of catalytic redox-active cysteines in thiol oxidoreductases. *Molecules and Cells*, 26: 228-235.

Forman, H. J., Maiorino, M. and Ursini, F. (2010) Signaling functions of reactive oxygen species. *Biochemistry*, 49: 835-842.

Franciscato, C., Moraes-Silva, L., Duarte, F. A., Oliveira, C. S., Ineu, R. P., Flores, E. M. M., Dressler, V. L., Peixoto, N. C. and Pereira, M. E. (2011) Delayed biochemical changes induced by Mercury intoxication are prevented by zinc pre-exposure. *Ecotoxicology and Environmental Safety*, 74: 480-486.

Frasco, M. F., Colletier, J-P., Weik, M., Carvalho, F., Guilhermino, L., Stojan, J. and Fournier, D. (2007) Mechanism of cholinesterase inhibition by inorganic mercury. *The FEBS Journal*, 274: 1849-1861.

Fritz-Wolf, K., Urig, S. and Becker, K. (2007) The structure of human thioredoxin reductase 1 provides insights into C-terminal rearrangements during catalysis. *The Journal of Molecular Biology*, 370: 116-127.

Frustaci, A., Neri, M., Cesario, A., Adams, J. B., Domenici, E., Dalla Bernardina, B. and Bonassi, S. (2012) Oxidative stress-related biomarkers in autism: Systematic review and meta-analyses. *Free Radical Biology and Medicine*, 52: 2128-2141.

Fuhr, B. J. and Rabenstein, D. L. (1973) Nuclear magnetic resonance studies of the solution chemistry of metal complexes. IX. The binding of cadmium, zinc, lead, and mercury by glutathione. *Journal of the American Chemical Society,* 95: 6944-6950.

Galembeck, S. E. and Caramori, G. F. (2003) Which is the reaction site? A computational experimente. *Química Nova*, 26: 957-959.

Giangregorio, N., Tonazzi, A., Console, L., Galluccio, M., Porcelli, V. and Indiveri, C. (2018) Structure/function relationships of the human mitochondrial ornithine/citrulline carrier by Cys site-directed mutagenesis. Relevance to mercury toxicity. *International Journal of Biological Macromolecules*, 120: 93-99.

Gilmour, C. C., Podar, M., Bullock, A. L., Graham, A. M., Brown, S. D., Somenahally, A. C., Johs, A., Hurt, R. A. Jr., Bailey, K. L. and Elias, D. A. (2013) Mercury methylation by novel microorganisms from new environments. *Environmental Science and Technology*, 47: 11810-11820.

Giustarini, D., Galvagni, F., Tesei, A., Farolfi, A., Zanoni, M., Pignatta, S., Milzani, A., Marone, I. M., Dalle-Donne, I., Nassini, R. and Rossi, R. (2015). Glutathione, glutathione disulfide, and S-glutathionylated proteins in cell cultures. *Free Radical Biology and Medicine*, 89: 972-981.

Giustarini, D., Lorenzini, S., Rossi, R., Chindamo, D., Di Simplicio, P. and Marcolongo, R. (2005). Altered thiol pattern in plasma of subjects affected by rheumatoid arthritis. *Clinical and Experimental Rheumatology*, 23: 205-212.

Glaser, V., Leipnitz, G., Straliotto, M. R., Oliveira, J., Santos, V. V., Wannmacher, C. M. D., de Bem, A. F., Rocha, J. B. T., Farina, M. and Latini, A. (2010) Oxidative stress-mediated inhibition of brain creatine kinase activity by methylmercury. *Neurotoxicology*, 31: 454-460.

Goldwater, L. J. (1965) The birth of mercaptan. *Archives of Environmental Health: An International Journal*, 11: 597-597.

Gorąca, A., Huk-Kolega, H., Piechota, A., Kleniewska, P., Ciejka, E. and Skibska, B. (2011) Lipoic acid – biological activity and therapeutic potential. *Pharmacological Reports*, 63: 849-858.

Grotto, D., Valentini, J., Fillion, M., Passos, C. J. S., Garcia, S. C., Mergler, D. and Barbosa Jr., F. (2010) Mercury exposure and oxidative stress in communities of the Brazilian Amazon. *Science of the Total Environment*, 408: 806-811.

Gusarov, I. and Nudler, E. (2018) Protein S-nitrosylation: Enzymatically controlled, but intrinsically unstable, post-translational modification. *Molecular Cell*, 69: 351-353.

Haouzi, D., Lekehal, M., Tinel. M., Vadrot. N., Caussanel, L., Lettéron, P., Moreau, A., Feldmann, G., Fau, D. and Pessayre, D. (2001) Prolonged, but not acute, glutathione depletion promotes Fas-mediated mitochondrial permeability transition and apoptosis mice. *Hepatology*, 33: 1181-1188.

Harada, M. (1995) Minamata disease: methylmercury poisoning in Japan caused by environmental pollution. *Critical Reviews in Toxicology* 25: 1-24.

Harris, T. K. and Turner, G. J. (2002) Structural basis of perturbed pKa values of catalytic groups in enzyme active sites. *IUBMB Life*, 53: 85-98.

Hartwig, A. (2001) Zinc fingers proteins as potential targets for toxic metal ions: Differential effects on structure and function. *Antioxidants & Redox Signaling*, 3: 625-634.

Hatton, L. and Warr, G. (2015) Protein structure and evolution: Are they constrained globally by a principle derived from information theory? *Plos One*, doi: 10.1371/journal.pone.0125663

He, F., Wang, W., Moon, J-W., Howe, J., Pierce, E. M. and Liang, L. (2012) Rapid removal of Hg(II) from aqueous solutions using thiol-functionalized Zn-doped biomagnetite particles. *Applied Material and Interfaces*, 4: 4373-4379.

Helms, C. and Shapiro, D. B. K. (2013) Hemoglobin-mediated nitric oxide signaling. *Free Radical Biology and Medicine*, 61:464-72.

Hernberg, S., Nikkanen, J. and Häsänen, E. (1971) Erythrocyte δ-aminolevulinic acid dehydratase in workers exposed to mercury vapor. *Scandinavian Journal of Work, Environment and Health*, 8: 42-45.

Ho, Y. B., Abdullah, N. H., Hamsan, H. and Tan, E. S. S. (2017) Mercury contamination in facial skin lightening creams and its health risks to user. *Regulatory Toxicology and Pharmacology*, 88: 72-76.

Holm, R. H., Kennepohl, P. and Solomon, E. I. (1996) Structural and functional aspects of metal sites in biology. *Chemical Reviews*, 96: 2239-2314.

Horowitz, H. M., Jacob, D. J., Amos, H. M., Streets, D. G. and Sunderland, E. M. (2014) Historical mercury releases from commercial products: global environmental implications. *Environmental Science and Technology*, 48: 10242-10250.

Hsu-Kim, H., Eckley, C. S. and Selin, N. E. (2018) Modern science of a legacy problem: mercury biogeochemical research after the Minamata Convention. *Environmental Science: Processes and Impacts*, 20: 582-583.

Huggins, F. E., Raverty, S. A., Nielsen, O. S., Sharp, N. E., Robertson, J. D. and Ralston, N. V. C. (2009) An XAFS investigation of mercury and selenium in beluga whale tissues. *Environmental Bioindicators*, 4: 291-302.

Ibne-Rasa, K. M. (1967). Equations for correlation of nucleophilic reactivity. *Journal of Chemical Education*, 44: 89-89.

Ihara, H., Kasamatsu, S., Kitamura, A., Nishimura, A., Tsutsuki, H., Ida, T., Ishizaki, K., Toyama, T., Yoshida, E., Abdul Hamid, H., Jung, M., Matsunaga, T., Fujii, S., Sawa, T., Nishida, M., Kumagai, Y. and Akaike, T. (2017) Exposure to electrophiles impairs reactive persulfide-dependent redox signaling in neuronal cells. *Chemical Research Toxicology*, 30: 1673-1684.

Infusino, I. and Panteghini, M. (2013) Serum albumin: Accuracy and clinical use. *Clinica Chimica Acta*, 419: 15-18.

Iqbalsyah, T. M., Velis, E. M., Warwicker, J., Errington, N. and Doig, A. J. (2006) The CXXC motif at the N-terminus of an α-helical peptide. *Protein Science*, 15: 1945-1950.

Ishitobi, H., Stern, S., Thurston, S. W., Zareba, G., Langdon, M., Gelein, R. and Weiss, B. (2010) Organic and inorganic mercury in neonatal rat brain after prenatal exposure to methylmercury and mercury vapor. *Environmental Health Perspectives*, 118: 242-248.

Jackson, A. C. (2018) Chronic neurological disease due to methylmercury poisoning. *The Canadian Journal of Neurological Sciences*, 0: 1-4.

Jagoe, C. H., Shaw-Allen, P. L. and Brundage S., (1996) Gill Na+,K+-ATPase activity in largemouth bass *(Micropterus salmoides)* from three reservoirs with different levels of mercury contamination. *Aquatic Toxicology*, 36: 161-176.

Jensen, K. A. (1989) The early history of organic sulfur chemistry. *Centaurus*, 32: 324-335.

Johansson, C., Smee, C., Kavanagh, K. L., Debreczeni, J., von Delft, F., Gileadi, O. et al. (2006) *Glutaredoxin 2 complexed with glutathione.* doi:10.2210/pdb2FLS/pdb.

Kabil, O. and Banerjee, R. (2010) Redox biochemistry of hydrogen sulfide. *The Journal of Biological Chemistry*, 285: 21903-21907.

Kanda, F., Kamikashi, T. and Ishibashi, S. (1976) Competitive inhibition of hexokinase isoenzymes by mercurials. *Journal of Biochemistry*, 79: 543-548.

Karplus, P. A. and Schulz, G. E. (1989) Substrate binding and catalysis by glutathione reductase as derived from refined enzyme: substrate crystal structures at 2 A resolution. *The Journal of Molecular Biology*, 210: 163-180.

Kataoka, H. (1998) Chromatographic analysis of lipoic acid and related compounds. *Journal of Chromatography B*, 717: 247-262.

Kehrig, H. A., Seixas, T. G., Palermo, E. A., Di Beneditto, A. P. M., Souza, C. M. M. and Malm, O. (2008) Different species of mercury in the livers of tropical dolphins. *Analytical Letters*, 41: 1691-1699.

Kerr, B. T., Ochs-Balcom, H. M., López, P., García-Vargas, G. G., Rosado, J. L., Cebrián, M. E. and Kordas, K. (2019). Effects of ALAD genotype on the relationship between lead exposure and anthropometry in a Cohort of Mexican children. *Environmental Research*, 170: 65-72.

Khan, M. A. K. and Wang, F. (2010) Chemical demethylation of methylmercury by selenoamino acids. *Chemical Research Toxicology*, 23: 1202-1206.

Kleinman, W. A. and Richie, J. P. (2000) Status of glutathione and other thiols and disulfides in human plasma. *Biochemical Pharmacology*, 60: 19-29.

Klomsiri, C., Karplus, P. A. and Poole, L. B. (2011) Cysteine-based redox switches in enzymes. *Antioxidants and Redox signaling*, 14: 1065-1077.

Klug, A. and Schwabe, J. W. (1995) Protein motifs 5. Zinc fingers. *FASEB Journal*, 9: 597-604.

Knerlich-Lukoschus, F. and Held-Feindt, J. (2015) Chemokine-ligands/receptors: multiplayers in traumatic spinal cord injury. *Mediators of Inflammation*, doi: 10.1155/2015/486758.

Kreimer, D. I., Dolginova, E. A., Raves, M., Sussman, J. L., Silman, I. and Weiner L. (1994) A metastable state of Torpedo californica acetylcholinesterase generated by modification with organomercurials. *Biochemistry*, 33: 14407-14418.

Krishna, S. S., Majumdar, I. and Grishin, N. V. (2003) Structural classification of zinc fingers. *Nucleic Acids Research*, 31: 532-550.

Kröncke, K-D. and Klotz, L. O. (2009) Zinc fingers as biologic redox switches. *Antioxidants and Redox Signaling*, 11: 1015-1027.

Kumar, D. and Banerjee, D. (2017) Methods of albumin estimation in clinical biochemistry: Past, present, and future. *Clinica Chimica Acta*, 469: 150-160.

Kuśmierek, K., Chwatko, G., Głowacki, R. and Bald, E. (2009) Determination of endogenous thiols and thiol drugs in urine by HPLC with ultraviolet detection. *Journal of Chromatography B*, 877: 3300-3308.

Laity, J. H., Lee, B. M. and Wright, P. E. (2001) Zinc finger proteins: new insights into structural and functional diversity. *Current Opinion in Structural Biology*, 11: 39-46.

Leiva-Presa, À., Capdevila, M. and Gonzàlez-Duarte, P. (2004) Mercury (II) binding to metallothioneins. Variables governing the formation and structural features of the mammalian Hg-MT species. *European Journal of Biochemistry*, 271: 4872-4880.

Lenz, G. R. and Martell A. E. (1964) Metal chelates of some sulfur-containing amino acids. *Biochemistry*, 3: 745-750.

Leonardi, R., Zhang, Y. M., Rock, C. O. and Jackowski, S. (2005) Coenzyme A: Back in action. *Progress in Lipid Research*, 44: 125-153.

Levy, E. J., Anderson, M. E. and Meister, A. (1993) Transport of glutathione diethyl ester into human cells. *Proceedings of the National Academy of Sciences of the United States of America*, 90: 9171-9175.

Li, F., Lutz, P. B., Pepelyayeva, Y., Arnér, E. S. J., Bayse, C. A. and Rozovsky, S. (2014) Redox active motifs in selenoproteins. *Proceedings of the National Academy of Sciences of the United States of America*, 111: 6976-6981.

Li, G., Zhao, Z., Liu, J. and Jiang, G. (2011) Effective heavy metal removal from aqueous systems by thiol functionalized magnetic mesoporous silica. *Journal of Hazardous Materials*, 192: 277-283.

Li, H., Robertson, A. D. and Jensen, J. H. (2005) Very fast empirical prediction and rationalization of protein pKa values. *Proteins*, 61: 704-721.

Li, X., Li, Z., Wu, T., Chen, J., Fu, C., Zhang, L., Feng, X., Fu, X., Tang, L., Wang, Z. and Wang Z. (2019) Atmospheric mercury emissions from two pre-calciner cement plants in Southwest China. *Atmospheric Environment*, 199: 177-188.

Li, Z., Jia, P., Zhao, Fu. and Kang, Y. (2018) Mercury pollution, treatment and solutions in spent fluorescent lamps in Mainland China. *International Journal of Environmental Research and Public Health*, doi:10.3390/ijerph15122766.

Limón-Pacheco, J. H., Hernández, N. A., Fanjul-Moles, M. L. and Gonsebatt, M. E. (2007) Glutathione depletion activates mitogen-activated protein kinase (MAPK) pathways that display organ specific responses and brain protection in mice. *Free radical Biology and Medicine*, 43: 1335-1347.

Lipman, D. J., Souvorov, A., Koonin, E. V., Panchenko, A. R. and Tatusova, T. A. (2002) The relationship of protein conservation and sequence length. *Evolutionary Biology*, doi: 10.1186/1471-2148-2-20.

Lippi, G., Mattiuzzi, C., Meschi, T., Cervellin, G. and Borghi, L. (2014) Homocysteine and migraine. A narrative review. *Clinica Chimica Acta*, 433: 5-11.

Liu, L., Wang, J., Wang, L., Hu, Y. and Ma, X. (2019). Vertical distributions of mercury in marine sediment cores from central and southern part of Bohai Sea, China. *Ecotoxicology and Environmental Safety*, 170: 399-406.

Liu, Y., Zhang, W., Zhao, J., Lin, X., Liu, J., Cui, L., Gao, Y., Zhang, T.-L., Li, B. and Li, Y.-F. (2018) Selenoprotein P as the major transporter for mercury in serum from methylmercury-poisoned rats. *Journal of Trace Elements in Medicine and Biology*, 50: 589-595.

Longo, A., Di Toro, M., Galimberti, C. and Carenzi, A. (1991) Determination of N-acetylcysteine in human plasma by gas chromatography-mass spectrometry. *Journal of Chromatography*, 562: 639-645.

LoPachin, R. M. and Gavin, T. (2014) Molecular mechanisms of aldehyde toxicity: a chemical perspective. *Chemical Research Toxicology*, 21: 1081-1091.

LoPachin, R. M., Gavin, T., DeCaprio, A. and Barber, D. S. (2012) Application of the Hard and Soft, Acids and Bases (HSAB) theory to toxicant - Target interactions, *Chemical Research Toxicology*, 25: 239-251.

Lu, J. and Holmgren, A. (2014) The thioredoxin antioxidant system. *Free Radical Biology and Medicine*, 66: 75-87.

Lu, J., Stewart, A. J., Sadler, P. J., Pinheiro, T. J. and Blindauer, C. A. (2008) Albumin as a zinc carrier: properties of its high-affinity zinc-binding site. *Biochemical Society Transactions*, 36: 1317-1321.

Lu, W., Zelazowski, A. J. and Stillman, M. J. (1993) Mercury binding to metallothioneins: formation of the Hg_{18}-MT species. *Inorganic Chemistry*, 32: 919-926.

Lyles, M. M. and Gilbert, H. F. (1991) Catalysis of the oxidative folding of ribonuclease A by protein disulfide isomerase: dependence of the rate on the composition of the redox buffer. *Biochemistry*, 30: 613-619.

Magos, L., Halbach, S. and Clarkson, T. W. (1978) role of catalase in the oxidation of mercury vapor. *Biochemical Pharmacology*, 27: 1373-1377.

Mansoor, M. A., Svardal, A. M. and Ueland, P. M. (1992) Determination of the *in vivo* redox status of cysteine, cysteinylglycine, homocysteine, and glutathione in human plasma. *Analytical Biochemistry*, 200: 218-229.

Manusadzianas, L., Maksimov, G., Rate, J., Iene, D., Jurkoniene, S., Sadauskas, K. and Vitkus, R. (2002) Response of the charophyte *Nitellopsis obtuse* to heavy metals at the cellular, cell membrane, and enzyme levels. *Environmental Toxicology*, 17: 275-283.

Maramba, N. P., Reyes, J. P., Francisco-Rivera, A. T., Panganiban, L. C. R., Dioquino, C., Dando, N., Timbang, R., Akagi, H., Castillo, M. T., Quitoriano, C., Afuang, M., Matsuyama, A., Eguchi, T. and Fuchigami, Y. (2006) Environmental and human exposure assessment monitoring of communities near an abandoned mercury mine in the Philippines: A toxic legacy. *Journal of Environmental Management*, 81: 135-145.

Maret, W. (2006) Zinc coordination environments in proteins as redox sensors and signal transducers. *Antioxidants and Redox Signaling*, 8: 1419-1441.

Martín-Doimeadios, R. C. R., Mateo, R. and Jiménez-Moreno, M. (2017) Is gastrointestinal microbiota relevant for endogenous mercury methylation in terrestrial animals? *Environmental Research*, 152: 454-461.

Masuda, T., Nakano, S. and Kondo, M. (1973) Rate constants for the reaction of OH radicals with the enzyme proteins as determined by the p-nitrosodimethylaniline method. *Journal of Radiation Research*, 14: 339-345.

Masutani, H., Ueda, S. and Yodoi, J. (2005) The thioredoxin system in retroviral infection and apoptosis. *Cell Death and Differentiation*, 12: 991-998.

Mattson, M. P. and Shea, T. B. (2003) Folate and homocysteine metabolism in neural plasticity and neurodegenerative disorders. *Trends in Neurosciences*, 26: 137-146.

Mavridou, D. A. I., Saridakis, E., Kritsiligkou, P., Mozley, E., Ferguson, S. J. and Redfield, C. (2014) An extended active-site motif controls the reactivity of the thioredoxin fold. *The Journal of Biological Chemistry*, 289: 8681-8696.

McBean, G. J. (2017) Cysteine, glutathione, and thiol redox balance in astrocytes. *Antioxidants*, doi: 10.3390/antiox6030062.

McBean, G. J., Aslan, M., Griffiths, H. R. and Torrão, R. C. (2015) Thiol redox homeostasis in neurodegenerative disease. *Redox Biology*, 5: 186-194.

McMenamin, M. E., Himmelfarb, J. and Nolin, T. D. (2009) Simultaneous analysis of multiple aminothiols in human plasma by high performance liquid chromatography with fluorescence detection. *Journal of Chromatography B*, 877: 3274-3281.

Melnick, J. G., Yurkerwich, K. and Parkin, G. (2010) On the chalcogenophilicity of mercury: evidence for strong Hg-Se bond in [Tm^{But}]HgSePh and its relevance to the toxicity of mercury. *Journal of the American Chemical Society,* 132: 647-655.

Merle, J.; Mazlo, J.; Watts, J., Moreno, R. and Ngu-Schwemlein, M. (2018) Reaction mixture analysis by ESI-MS: Mercury(II) and dicysteinyl tripeptide complex formation. *International Journal of Mass Spectrometry*, 426: 38-47.

Mesterházy, E., Lebrun, C., Crouzy, S., Jancsó, A. and Delangle, P. (2018) Short oligopeptides with three cysteine residues as models of sulphur-rich Cu(I)- and Hg(II)-binding sites in proteins. *Metallomics*, 10: 1232-1244.

Metanis, N., Keinan, E. and Dawson, P. E. (2006) Synthetic seleno-glutaredoxin 3 analogues are highly reducing oxidoreductases with enhanced catalytic efficiency. *Journal of the American Chemical Society,* 128: 16684-16691.

Mezyk, S. P. (1996a) Determination of the rate constant for the reaction of hydroxyl and oxide radicals with cysteine in aqueous solution. *Radiation Research*, 145: 102-106.

Mezyk, S. P. (1996b) Rate constant determination for the reaction of hydroxyl and glutathione thiyl radicals with glutathione in aqueous solution. *The Journal of Physical Chemistry*, 100: 8861-8866.

Mills-Davies, N., Butler, D., Norton, E., Thompson, D., Sarwar, M., Guo, J., Gill, R., Azim, N., Coker, A., Wood, S. P., Erskine, P. T., Coates, L., Cooper, J. B., Rashid, N., Akhtar, M. and Shoolingin-Jordan, P. M. (2017) Structural studies of substrate and product complexes of 5-aminolaevulinic acid dehydratase from humans, *Escherichia coli* and the hyperthermophile *Pyrobaculum calidifontis*. *Acta Crystallographica D*, 1: 9-21.

Monostori, P., Wittmann, G., Karg, E. and Túri, S. (2009) Determination of glutathione and glutathione disulfide in biological samples: An in-depth review. *Journal of Chromatography B*, 877: 3331-3346.

Morris, A. D., Letcher, R. J., Dyck, M., Chandramouli, B. and Cosgrove, J. (2019). Concentrations of legacy and new contaminants are related to metabolite profiles in Hudson Bay polar bears. *Environmental Research*, 168: 364-374.

Müller, S., Senn, H., Gsell, B., Vetter, W., Baron, C. and Böck, A. (1994) The formation of diselenide bridges in proteins by incorporation of selenocysteine residues: biosynthesis and characterization of (Se)2-thioredoxin. *Biochemistry*, 33: 3404-3412.

Muntean, M., Janssens-Maenhout, G., Song, S., Giang, A., Selin, N. E., Zhong, H., Zhao, Y., Olivier, J. G. J., Guizzardi, D., Crippa, M., Schaaf, E. and Dentener, F. (2018) Evaluating EDGARv4.tox2 speciated mercury emissions ex-post scenarios and their impacts on modeled global and regional wet deposition patterns. *Atmospheric Environment*, 184: 56-68.

Muntean, M., Janssens-Maenhout, G., Song, S., Selin, N. E., Olivier, J. G. J., Guizzardi, D., Maas, R. and Dentener, F. (2014) Trend analysis from 1970 to 2008 and model evaluation of EDGARv4 global gridded anthropogenic mercury emissions. *Science of the Total Environment*, 494-495: 337-350.

Nakamura, Y. K., Dubick, M. A. and Omaye, S. T. (2012) γ-Glutamylcysteine inhibits oxidative stress in human endothelial cells. *Life Sciences*, 90: 116-121.

Navari-Izzo, F., Quartacci, M. F. and Sgherri, C. (2002) Lipoic acid: a unique antioxidant in the detoxification of activated oxygen species. *Plant Physiology and Biochemistry*, 40: 463-470.

Netto, L. E. S., Oliveira, M. A., Monteiro, G., Demasi, A. P. D., Cussiol, J. R. R., Discola, K. F., Demasi, M., Silva, G. M., Alves, S. V., Faria, V. G. and Horta, B. B. (2007) Reactive cysteine in proteins: protein folding, antioxidante defense, redox signaling and more. *Comparative Biochemistry and Physiology C*, 146: 180-193.

Newman, W. R. (2014) Mercury and sulphur among the high medieval alchemists: from Razi and Avicenna to Albertus Magnus and Pseudo-Roger Bacon. *Ambix*, 61: 327-344.

Nicholson, D. W. and Thornberry, N. A. (1997) Caspases: killer proteases. *Trends in Biochemical Sciences*, 22: 299-306.

Nogara, P. A., Oliveira, C. S., Schmitz, G. L., Piquini, P. C., Farina, M. Aschner, M. and Rocha, J. B. T. (2019) Methylmercury's chemistry: from the environment to the mammalian brain. *Biochimica et Biophysica Acta - General Subjects*, doi: 10.1016/j.bbagen.2019. 01.006.

Nolin, T. D., McMenamin, M. E. and Himmelfarb, J. (2007) Simultaneous determination of total homocysteine, cysteine, cysteinylglycine, and glutathione in human plasma by high-performance liquid chromatography: application to studies of oxidative stress. *Journal of Chromatography B*, 852: 554-561.

Oakley, A. J., Lo Bello, M., Battistoni, A., Ricci, G., Rossjohn, J., Villar, H. O. and Parker, M. W. (1997) The structures of human glutathione transferase P1-1 in complex with glutathione and various inhibitors at high resolution. *The Journal of Molecular Biology*, 274: 84-100.

Ogata, M., Kenmotsu, K., Hirota, N., Meguro, T. and Aikoh, H. (1987) Reduction of mercuric ion and exhalation of mercury in acatalasemic and normal mice. *Archives of Environmental Health: An International Journal*, 42: 26-30.

Oliveira, C. S., Joshee, L. and Bridges, C. C. (2018b) MRP2 and the transport kinetics of cysteine conjugates of inorganic mercury. *Biological Trace Element Research*, 184: 279-286

Oliveira, C. S., Nogara, P. A., Ardisson-Araújo, D. M. P., Aschner, M., Rocha, J. B. T. and Dórea, J. G. (2018a) Neurodevelopmental effects of mercury. In: Aschner, M. and Costa, L. (Eds) *Linking environmental exposure to neurodevelopmental disorders. Advances in Neurotoxicology*, vol 2. Elsevier.

Oliveira, C. S., Piccoli, B. C., Aschner, M. and Rocha, J. B. T. (2017) Chemical speciation of selenium and mercury as determinant of their neurotoxicity. In: Aschner, M. and Costa, L. (Eds) *Neurotoxicity of Metals. Advances in Neurobiology*, vol. 18. Springer, Cham.

Olsen, M., Fjeld, E. and Lydersen, E. (2019) The influence of a submerged meadow on uptake and trophic transfer of legacy mercury from contaminated sediment in the food web in a brackish Norwegian fjord. *Science of the Total Environment*, 654: 209-217.

Onosaka, S., Min, K., Fukuhara, C., Tanaka, K., Tashiro, S., Furuta, M., Shimizu, I. and Yasutomi, T. (1985) Concentration of metallothionein in human tissues. *Eisei kagaku*, 31: 352-355.

Onosaka, S., Min, K.-S., Fukuhara, C., Tanaka, K., Tashiro, S.-I., Shimizu, I., Furuta, M., Yasutomi, T., Kobashi, K. and Yamamoto, K.-I. (1986) Concentrations of metallothionein and metals in malignant and non-malignant tissues in human liver. *Toxicology*, 38: 261-268.

Oram, P. D., Fang, X., Fernando, Q., Letkeman, P. and Letkeman, D. (1996) The formation of constants of mercury(II)-glutathione complexes. *Chemical Research in Toxicology*, 9: 709-712.

Parks, J. M., Johs, A., Podar, M., Bridou, R., Hurt, R. A. Jr., Smith, S. D., Tomanicek, S. J., Qian, Y., Brown, S. D., Brandt, C. C., Palumbo, A. V., Smith, J. C., Wall, J. D., Elias, D. A. and Liang, L. (2013) The genetic basis for bacterial mercury methylation. *Science*, 339: 1332-1335.

Paul, B. D., Sbodio, J. I. and Snyder, S. H. (2018) Cysteine metabolism in neuronal redox homeostasis. *Trends in Pharmacological Sciences*, 39: 513-524.

Pearson, R. G. (1963) Hard and Soft Acids and Bases. *Journal of the American Chemical Society*, 85: 3533-3539.

Pearson, R. G. (1990) Hard and soft acids and bases-the evolution of a chemical concept. *Coordination Chemistry Reviews* 100: 403-425.

Peixoto, N, C., Roza, T. and Pereira, M. E. (2004) Sensitivity of delta-ALA-D (E.C. 4.2.1.24) of rats to metals *in vitro* depends on the stage of postnatal growth and tissue. *Toxicol In Vitro*, 18: 805-809.

Pelham, H. R. B. and Brown, D. D. (1980) A specific transcription factor that can bind either the 5S RNA gene or 5S RNA. *Proceedings of the National Academy of Sciences of the United States of America*, 77: 4170-4174.

Perry, T. L. and Hansen, S. (1981) Cystinylglycine in plasma: diagnostic relevance for pyroglutamic acidemia, homocystinuria, and phenylketonuria. *Clinica Chimica Acta*, 117: 7-12.

Petitpas, I., Bhattacharya, A. A., Twine, S., East, M. and Curry, S. (2001) Crystal structure analysis of warfarin binding to human serum albumin: anatomy of drug site I. *The Journal of Biological Chemistry*, 276: 22804-22809.

Pietrocola, F., Galluzzi, L., Bravo-San Pedro, J. M., Madeo, F. and Kroemer, G. (2015) Acetyl coenzyme A: A central metabolite and second messenger. *Cell Metabolism*, 21: 805-821.

Pisoschi, A. M. and Pop, A. (2015) The role of antioxidants in the chemistry of oxidative stress: A review. *European Journal of Medicinal Chemistry*, 97: 55-74.

Poole, L. B. (2015) The basics of thiols and cysteines in redox biology and chemistry. *Free Radical Biology and Medicine*, 80: 148-157.

Prakash, M., Shetty, M. S., Tilak, P. and Anwar, N. (2009) Total thiols: Biomedical importance and their alteration in various disorders. *The Online Journal of Health and Allied Sciences*, 8: 1-9.

Quijano, C., Alvarez, B., Gatti, R. M., Augusto, O. and Radi, R. (1997) Pathways of peroxynitrite oxidation of thiol groups. *Biochemical Journal*, 322: 167-173.

Rabeh, W. M., Zhu, H., Nedyalkova, L., Tempel, W., Wasney, G., Landry, R. et al. (2006) *Crystal structure of human hexokinase II.* doi:10.2210/pdb2NZT/pdb.

Rabenstein, D. L. (1978) The chemistry of methylmercury toxicity. *Journal of Chemical Education*, 37: 292-296.

Rabenstein, D. L. and Evans, C A. (1978) The mobility of methylmercury in biological systems. *Bioinorganic Chemistry*, 8: 107-114.

Rabenstein, D. L. and Reid, R. S. (1984) Nuclear magnetic resonance studies of the solution chemistry of metal complexes. 20. Ligand-exchange kinetics of methylmercury(II)-thiol complexes. *Inorganic Chemistry*, 23: 1246-1250.

Radi, R., Beckman, J. S., Bush, K. M. and Freeman, B. (1991) Peroxynitrite oxidation of sulfhydryls. The cytotoxic potential of superoxide and nitric oxide. *The Journal of Biological Chemistry*, 266: 4244-4250.

Ragunathan, N., Busi, F., Pluvinage, B., Sanfins, E., Dupret, J. M., Rodrigues-Lima, F. and Dairou, J. (2010) The human xenobiotic-metabolizing enzyme arylamine N-acetyltransferase 1 (NAT1) is irreversibly inhibited by inorganic (Hg^{2+}) and organic mercury (CH_3Hg^+): Mechanism and kinetics, *FEBS Letters*, 584: 3366-3369.

Rajanna, B., Chettyb, C. S., Rajanna, S., Hall, E., Fail, S. and Yallapragada, P. R. (1995) Modulation of protein kinase C by heavy metals. *Toxicology Letters*, 81: 197-203.

Reddie, K. G. and Carrol, K. S. (2008) Expanding the functional diversity of proteins through cysteine oxidation. *Current Opinion in Chemical Biology*, 12: 746-754.

Reid, R. S. and Rabenstein, D. L. (1981) Nuclear magnetic resonance studies of the solution chemistry of metal complexes. XVII. Formation constants for the complexation of methylmercury by sulfhydryl-containing amino acids and related molecules. *Canadian Journal of Chemistry*, 59:1505-1514.

Ren, X., Zou, L., Zhang, X., Branco, V., Wang, J., Carvalho, C., Holmgren, A. and Lu, J. (2017) Redox signaling mediated by thioredoxin and glutathione systems in the central nervous system. *Antioxidants and Redox Signaling*, 27: 989-1010.

Ritov, V. B. and Kelley, D. E. (2001) Hexokinase isozyme distribution in human skeletal muscle. *Diabetes*, 50: 1253-1262.

Rocha, J. B. T., Saraiva, R. A., Garcia, S. C., Gravina, F. S. and Nogueira, C. W. (2012) Aminolevulinate dehydratase (δ-ALA-D) as marker protein of intoxication with metals and other pro-oxidant situations. *Toxicology Research*, 1: 85-102.

Rodrigues, J., Branco, V., Lu, J., Holmgren, A. and Carvalho, C. (2015) Toxicological effects of thiomersal and ethylmercury: Inhibition of the thioredoxin system and NADP$^+$-dependent dehydrogenases of the pentose phosphate pathway. *Toxicology and Applied Pharmacology*, 286: 216-223.

Rooney, J. P. K. (2007) The role of thiols, dithiols, nutritional factors and interactings ligands in the toxicology of mercury. *Toxicology*, 234: 145-156.

Roos, D. H., Puntel, R. L., Farina, M., Aschner, M. Bohrer, D., Rocha, J. B. T. and Barbosa, N. B. V. (2011) Modulation of methylmercury uptake by methionine: Prevention of mitochondrial dysfunction in rat liver slices by a mimicry mechanism. *Toxicology and Applied Pharmacology*, 252: 28-35.

Rudd, J. W., Bodaly, R. A., Fisher, N. S., Kelly, C. A., Kopec, D. and Whipple, C. (2018) Fifty years after its discharge, methylation of legacy mercury trapped in the Penobscot Estuary sustains high mercury in biota. *Science of the Total Environment*, 642: 1340-1352.

Sakamoto, M., Itai, T., Yasutake, A., Iwasaki, T., Yasunaga, G., Fujise, Y., Nakamura, M., Murata, K., Chan, H. M., Domingo, J. L. and Marumoto, M. (2015) Mercury speciation and selenium in toothed-whale muscles. *Environmental Research*, 143: 55-61.

Sayın, D., Çakır, D. T., Gençer, N. and Arslan O. (2012) Effects of some metals on paraoxonase activity from shark *Scyliorhinus canícula*. *Journal of Enzyme Inhibition and Medicinal Chemistry*, 27: 595-598.

Schütz, A. and Skerfving, S. (1975) Blood cell δ-aminolevulinic acid dehydratase activity in humans exposed to methylmercury. *Scandinavian Journal of Work, Environment and Health*, 1: 54-59.

Short, J. D., Downs, K., Tavakoli, S. and Asmis, R. (2016) Protein thiol redox signaling in monocytes and macrophages. *Antioxidants and Redox Signaling*, 25: 816-835.

Shurubor, Y. I., D'Aurelio, M., Clark-Matott, J., Isakova, E. P., Deryabina, Y. I., Beal, M. F., Cooper, A. J. L. and Krasnikov, B. F. (2017) Determination of coenzyme A and acetyl-coenzyme A in biological samples using HPLC with UV detection. *Molecules*, doi: 10.3390/molecules22091388.

Sies, H. (1991) Oxidative stress: from basic research to clinical application. *The American Journal of Medicine*, 91: 31-38.

Siltberg-Liberles, J., Grahnen, J. A. and Liberles, D. A. (2011) The evolution of protein structures and structural ensembles under functional constraint. *Genes*, 2: 748-762.

Silva, K. C. C., Assis, C. R. D., Oliveira, V. M., Carvalho, Jr, L. B. and Bezerra, R. S. (2013) Kinetic and physicochemical properties of brain acetylcholinesterase from the peacock bass (*Cichla ocellaris*) and *in vitro* effect of pesticides and metal ions. *Aquatic Toxicology*, 126: 191-197.

Smith, J. C., Farris, F. F. and Burg, R. (1976) Chemical type of mercury in patients in the outbreak of organomercury poisoning in Iraq. *Bulletin of the World Health Organization*, 53: 61-63.

Soares Moretti, A. I. and Martins Laurindo, F. R. (2017) Protein disulfide isomerases: Redox connections in and out of the endoplasmic reticulum. *Archives of Biochemistry and Biophysics* 617: 106-119.

Streets, D. G., Horowitz, H. M., Jacob, D. J., Lu, Z., Levin, L., Ter Schure, A. F. and Sunderland, E. M. (2017) Total mercury released to the environment by human activities. *Environmental Science and Technology*, 51: 5969-5977.

Streets, D. G., Horowitz, H. M., Lu, Z., Levin, L., Thackray, C. P. and Sunderland, E. M. (2018) Global and regional trends in mercury emissions and concentrations, 2010-2015. *Atmospheric Environment*, 201: 417-427.

Stricks, W. and Kolthoff, I. M. (1953) Reactions between mercuric mercury and cysteine and glutathione. Apparent dissociation constants, heats and entropies of formation of various forms of mercuric mercapto-cysteine and -glutathione. *Journal of the American Chemical Society*, 75: 5673-5681.

Sueishi, Y., Hori, M., Ishikawa, M., Matsu-ura, K., Kamogawa, E., Honda, Y., Kita, M. and Ohara, K. (2014) Scavenging rate constants of hydrophilic antioxidants against multiple reactive oxygen species. *Journal of Clinical Biochemistry and Nutrition*, 54: 67-74.

Sun, K., D'Alessandro, A., Ahmed, M. H., Zhang, Y., Song, A., Ko, T. P., Nemkov, T., Reisz, J. A., Wu, H., Adebiyi, M., Peng, Z., Gong, J., Liu, H., Huang, A., Wen, Y. E., Wen, A. Q., Berka, V., Bogdanov, M. V., Abdulmalik, O., Han, L., Tsai, A. L., Idowu, M., Juneja, H. S., Kellems, R. E., Dowhan, W., Hansen, K. C., Safo, M. K. and Xia Y. (2017) Structural and functional insight of sphingosine 1-phosphate-mediated pathogenic metabolic reprogramming in sickle cell disease. *Scientific Reports*, doi: 10.1038/s41598-017-13667-8.

Sutherland, D. E. K. and Stillman, M. J. (2011) The "magic numbers" of metallothionein. *Metallomics*, 3: 444-463.

Szczurek, E. I., Taylor, C. G. and Bjornsson, C. S. (2001) Dietary zinc deficiency and repletion modulate metallothionein immunolocalization and concentration in small intestine and liver of rats. *The Journal of Nutrition*, 131: 2132-2138.

Tasmin, S., Furusawa, H., Ahmad, S. A., Faruquee, M. H. and Watanabe, C. (2015) Delta-aminolevulinic acid dehydratase (ALAD) polymorphism in lead exposed Bangladeshi children and its effect on urinary aminolevulinic acid (ALA). *Environmental Research*, 136: 318-323.

Tekaya, N., Saiapina, O., Ouada, H. B., Lagarde, F., Ouada, H. B. and Jaffrezic-Renault, N. (2013) Ultra-sensitive conductometric detection of heavy metals based on inhibition of alkaline phosphatase activity from *Arthrospira platensis*, *Bioelectrochemistry*, 90: 24-29.

Tekman, B., Ozdemir, H., Senturk, M. and Ciftci, M. (2008) Purification and characterization of glutathione reductase from rainbow trout (*Oncorhynchus mykiss*) liver and inhibition effects of metal ions on enzyme activity. *Comparative Biochemistry and Physiology C*, 148: 117-121.

Thanan, R., Oikawa, S., Hiraku, Y., Ohnishi, S., Ma, N., Pinlaor, S., Yongvanit, P., Kawanish, S. and Murata, M. (2015) Oxidative stress ans its significant roles in neurodegenerative diseases and cancer. *International Journal of Molecular Science*, 16: 193-217.

Thirumoorthy, N., Manisenthil Kumar, K. T., Shyam Sundar, A., Panayappan, L. and Chatterjee, M. (2007) Metallothionein: an overview. *World Journal of Gastroenterology*, 13: 993-996.

Tinkov, A. A., Ajsuvakova, O. P., Skalnaya, M. G., Popova, E. V., Sinitskii, A. I., Nemereshina, O. N., Gatiatulina, E. R., Nikonorov, A. A. and Skalny, A. V. (2015) Mercury and metabolic syndrome: a review of experimental and clinical observations. *Biometals*, 28: 231-254.

Tonazzi, A., Giangregorio, N., Console, L., Scalise, M., La Russa, D., Notaristefano, C., Brunelli, E., Barca, D. and Indiveri C. (2015) Mitochondrial carnitine/acylcarnitine transporter, a novel target of mercury toxicity. *Chemical Research Toxicology*, 28: 1015−1022.

Torres, M. J., Turell, L., Botti, H., Antmann, L., Carballal, S., Ferrer-Sueta, G., Radi, R. and Alvarez, B. (2012) Modulation of the reactivity of the thiol of human serum albumin and its sulfenic derivative by fatty acids. *Archives of Biochemistry and Biophysics*, 521: 102-110.

Trujillo, M. and Radi, R. (2002) Peroxynitrite reaction with the reduced and the oxidized forms of lipoic acid: New insights into the reaction of peroxynitrite with thiols. *Archives of Biochemistry and Biophysics*, 397: 91-98.

Turell, L., Radi, R. and Alvarez, B. (2013) The thiol pool in human plasma: the central contribution of albumin to redox processes. *Free Radical Biology and Medicine*, 65: 244-253.

Vahter, M., Mottet, N. K., Friberg, L., Lind, B., Shen, D. D. and Burbacher, T. (1994) Speciation of mercury in the primate blood and brain following long-term exposure to methylmercury. *Toxicology and Applied Pharmacology*, 124: 221-229.

Vazquez-Torrez, A. (2012) Redox active thiol sensors of oxidative and nitrosative stress. *Antioxidants and Redox Signaling*, 17: 1201-1214.

Veiga, M. M. and Hinton, J. J. (2002) Abandoned artisanal gold mines in the Brazilian Amazon: a legacy of mercury pollution. In *Natural resources forum*, 26: 15-26. Oxford, UK and Boston, USA: Blackwell Publishing Ltda.

Wagemann, R., Trebacz, E., Boila, G. and Lockhart, W. L. (2000) Mercury species in the liver of ringed seals. T*he Science of the Total Environment,* 261: 21-32.

Wagner, C., Sudati, J. H., Nogueira, C. W. and Rocha, J. B. T. (2010) *In vivo* and *in vitro* inhibition of mice thioredoxin reductase by methylmercury. *Biometals*, 23: 1171-1177.

Walsh, C. T., Garneau-Tsodikova, S. and Gatto, G. J. Jr. (2005) Protein posttranslational modifications: the chemistry of proteome diversifications. *Angewandte Chemie*, 44: 7342-7372.

Walsh, M. J. and Ahner, B. A. (2013) Determination of stability constants of Cu(I), Cd(II) and Zn(II) complexes with thiols using fluorescent probes. *Journal of Inorganic Biochemistry*, 128: 112-123.

Wang, C., Li, W., Ren, J., Fang, J., Ke, H., Gong, W., Feng, W. and Wang, C. C. (2013) Structural insights into the redox-regulated dynamic conformations of human protein disulfide isomerase. *Antioxidants and Redox Signaling*, 19: 36-45.

Wang, S., Lv, Q., Yang, Y., Guo, L-H., Wan, B., Ren, X. and Zhang H. (2016) Arginine decarboxylase: A novel biological target of mercury compounds identified in PC12 cells. *Biochemical Pharmacology*, 118: 109-120.

Wardman, P., Dennis, M. F., Stratford, M. R. and White, J. (1992) Extracellular: intracellular and subcellular concentration gradients of thiols. *International Journal of Radiation Oncology, Biology, Physics*, 22: 751-754.

Warren, M. J., Cooper, J. B., Wood, S. P. and Shoolingin-Jordan, P. M. (1998) Lead poisoning, haem synthesis and 5-aminolaevulinic acid dehydratase. *Trends in Biochemical Sciences*, 23: 217-221.

Watanabe, M. M., Laurindo, F. R. M. and Fernandes, D. C. (2014) Methods of measuring protein disulfide isomerase activity: a critical overview. *Frontiers in Chemistry*, 2: 73. doi: 10.3389/fchem.2014. 00073.

Weichsel, A., Gasdaska, J. R., Powis, G. and Montfort, W. R. (1996) Crystal structures of reduced, oxidized, and mutated human thioredoxins: evidence for a regulatory homodimer. *Structure*, 4: 735-751.

Winterbourn, C. C. and Hampton, M. B. (2008) Thiol chemistry and specificity in redox signaling. *Free Radical Biology and Medicine*, 45: 549-561.

Winterbourn, C. C. and Metodiewa, D. (1999) Reactivity of biologically important thiol compounds with superoxide and hydrogen peroxide. *Free Radical Biology and Medicine*, 27: 322-328.

Witkiewicz-Kucharczyk, A. and Bal, W. (2006) Damage of zinc fingers in DNA repair proteins, a novel molecular mechanism in carcinogenesis. *Toxicology Letters*, 162: 29-42.

Wouters, M. A., Fan, S. W. and Haworth, N. L. (2010) Disulfides as redox switches: from molecular mechanisms to functional significance. *Antioxidants and Redox Signaling*, 12: 53-91.

Wouters, M. A., George, R. A. and Haworth, N. L. (2007) "Forbidden" disulfides: their role as redox switches. *Current Protein and Peptide Science*, 8: 484-495.

Woycechowsky, K. J. and Raines, R. T. (2003) The CXC motif: A functional mimic of protein disulfide isomerase. *Biochemistry*, 42: 5387-5394.

Wu Y., Yang X., Zhang B. and Guo L. H. (2015) An electrochemiluminescence biosensor for 8-oxo-7,8-dihydro-2'-deoxyguanosine quantification and DNA repair enzyme activity analysis using a novel bifunctional probe. *Biosensors and Bioelectronics*, 69: 235-240.

Xu, J., Buck, M., Eklöf, K., Ahmed, O. O., Schaefer, J. K., Bishop, K., Skyllberg, U., Bertilsson, S. and Bravo, A. G. (2019). Mercury methylating microbial communities of boreal forest soils. *Scientific reports*, doi:10.1038/s41598-018-37383-z.

Xu, M., Yang, L. and Wang, Q. (2013) Chemical interactions of mercury species and some transition and noble metals towards metallothionein (Zn_7MT-2) evaluated using SEC/ICP-MS, RP-HPLC/ESI-MS and MALDI-TOF-MS. *Metallomics*, 5: 855-860.

Yang, Y., Li, L., Hang, Q., Fang, Y., Dong, X., Cao, P., Yin, Z. and Luo, L. (2019) γ-glutamylcysteine exhibits anti-inflammatory effects by increasing cellular glutathione level. *Redox Biology*, 20: 157-166.

Zalups, R. K. (1993) Early aspects of the intrarenal distribution of mercury after the intravenous administration of mercuric chloride. *Toxicology*, 79: 215-228.

Zalups, R. K. and Barfuss, D. W. (1993) Intrarenal distribution of inorganic mercury and albumin after coadministration. *Journal of Toxicology and Environmental Health*, 40: 77-103.

Zalups, R. K. and Barfuss, D. W. (1995a) Accumulation and handling of inorganic mercury in the kidney after coadministration with glutathione. *Journal of Toxicology and Environmental Health*, 44: 385-399.

Zalups, R. K. and Barfuss, D. W. (1995b) Renal disposition of mercury in rats after intravenous injection of inorganic mercury and cysteine. *Journal of Toxicology and Environmental Health*, 44: 401-413.

Zalups, R. K. and Barfuss, D. W. (1996) Nephrotoxicity of inorganic mercury co-administered with L-cysteine. *Toxicology*, 109: 15-29.

Zalups, R. K. and Bridges, C. C. (2009) MRP2 involvement in renal proximal tubular elimination of methylmercury mediated by DMPS or DMSA. *Toxicology and Applied Pharmacology*, 235: 10-17.

Zalups, R. K. and Lash, L. H. (2006) Cystine alters the renal and hepatic disposition of inorganic mercury and plasma thiol status. *Toxicology and Applied Pharmacology*, 214: 88-97.

Zeida, A., Guardia, C. M., Lichtig, P., Perissinotti, L. L., Defelipe, L. A., Turjanski, A., Radi, R., Trujillo, M. and Estrin, D. A. (2014) Thiol redox biochemistry: insights from computer simulations. *Biophysical Reviews*, 6: 27-46.

Zhang, Q., Sun, X., Sun, S., Yin, X., Huang, J., Cong, Z. and Kang, S. (2019). Understanding mercury cycling in Tibetan glacierized mountain environment: Recent progress and remaining gaps. *Bulletin of Environmental Contamination and Toxicology*, doi:10.1007/s00128-019-02541-0.

Zhang, Y., Tian, T. R. H., Jin, B. and He, J. (2018) Hydrogel-encapsulated enzyme facilitates colorimetric acute toxicity assessment of heavy metal ions. *Applied Materials and Interfaces*, 10: 26705-26712.

Ziller, A., Yadav, R. K., Capdevilla, M., Reddy, M. S., Vallon, L., Marmeisse, R., Atrian, S., Palacios, Ó. and Fraissinet-Tachet, L. (2017) Metagenomics analysis reveals a new metallothionein family: sequence and metal-binding features of new environmental cysteine-rich proteins. *Journal of Inorganic Biochemistry*, 167: 1-11.

Zimmermann, L. T., Santos, D. B., Naime, A. A., Leal, R. B., Dórea, J. G., Barbosa Jr, F., Aschner, M., Rocha, J. B. T. and Farina, M.(2013) Comparative study on methyl- and ethylmercury-induced toxicity in C6 glioma cells and the potential role of LAT-1 in mediating mercurial-thiol complexes uptake. *Neurotoxicology*, 38: 1-8.

In: Thiols: Structure, Properties and Reactions   ISBN: 978-1-53615-599-0
Editor: Carlos C. McAlpine   © 2019 Nova Science Publishers, Inc.

Chapter 2

THIOL-METHACRYLATE NETWORKS AND ITS APPLICATION IN NOVEL HYDROGELS

*Claudia I. Vallo** and Silvana V. Asmussen*

Institute of Materials Science and Technology (INTEMA),
University of Mar del Plata, CONICET, Mar del Plata, Argentina

ABSTRACT

Thiol-ene systems have attracted considerable recent interest because they display efficient, versatile and selective "click" reactions. In this study, thiol-methacrylate networks based on a tetrafunctional thiol (PTMP) and dimethacrylate monomers were prepared by both photopolymerization and amine-catalyzed Michael addition reaction. The progress of the polymerization reaction was monitored by FTIR and Raman spectroscopy. The cured materials were characterized by measuring the glass transition temperature, the flexural modulus and the compressive strength. Innovative hydrogels based on PTMP and a water-soluble dimethacrylate monomer (LBisEMA) were prepared by both visible light photo polymerization and Michael addition reaction. Depending on the synthesis

* Corresponding Author's E-mail: civallo@fi.mdp.edu.ar.

method employed significant differences in the degree of swelling were observed. LBisEMA–PTMP hydrogels prepared by the Michael addition reaction catalyzed by propylamine resulted in the highest water uptake. Hydrogels prepared by visible light photo polymerization contained unreacted thiol groups because of a faster homopolymerization reaction of the methacrylate groups. No significant effect of the PTMP/LBisEMA molar ratio on the water sorption capacity of the photocured hydrogel was observed. These trends are explained in terms of a balance between the mass fraction of hydrophilic groups and the crosslinking density of the network. Silver nanoparticles can be easily incorporated into thiol-ene systems because of the stabilizing properties of thiol functional groups. These polymer nanocomposites are very attractive for the preparation of biomaterials or dermatological patches with improved biocompatibility. Silver nanoparticles were prepared in PTMP by in situ reduction of silver nitrate with 2,6-di-tert-butyl-p-cresol. In this system, the thiol monomer acts as both stabilizing agent and reactive solvent. Mixtures of PTMP containing silver nanoparticles and a bifunctional methacrylate monomer were photoactivated with 2,2-dimethoxy-2-phenylacetophenone or camphorquinone (CQ) and then photopolymerized by irradiation with UV or visible light respectively. Visible light photo polymerization is commonly carried out using the CQ-amine photoinitiator system. CQ displays an intense dark yellow color which bleaches under irradiation giving as a result colorless polymers. However, we found that in thiol-ene systems CQ is regenerated through hydrogen transfer reactions between ketyl radicals and thiyl radicals. This feature must be taking into account when considering the long-term color stability of the derived cured materials.

Keywords: thiol-methacrylate, photopolymerization, hydrogels, silver nanoparticles

INTRODUCTION

Polymerization of thiol-ene systems displays many of the features of "click" reactions [1-4]. These features include rapid polymerization rates with reactions occurring either in environmentally benign solvents or in bulk, being insensitive to the presence of water or oxygen. This extraordinary versatility makes thiol-ene chemistry adequate for a wide range of applications [1-4]. Thiol-ene polymerization is initiated through the generation of radical centers, being photo induced initiation the most common method [1-3]. Radicals generated by irradiation of resins containing photoinitiator systems result in the formation of a thiyl radicals. Once formed, the thiyl radical initiates the polymerization by inserting into the double bond to give a carbon-centered radical, thus initiating the radical-chain process [1-4]. Termination occurs by radical-radical coupling. In addition to the photo induced free radical polymerization reaction, thiol-ene addition reactions occur through base- or nucleophile-catalyzed Michael addition reaction [5-6] Primary and secondary amines produce deprotonation of thiol groups thereby generating thiolate anion and ammonium cation. The thiolate anion adds at the β-carbon of the double bond producing an intermediate carbon-centered anion which is a very strong base. Once the enolate anion picks up a proton either from a thiol group or from the ammonium cation the thiol-ene product is formed. [3] In this report, the photopolymerization of thiol-methacrylate networks photoinitiated with Camphorquinone and 2,2-Dimethoxy-2-phenylace-tophenone is described. The Michael addition reaction catalyzed by propylamine is also presented.

Hydrogels have attracted much interest in recent years because they can be used in numerous applications such as soft contact lenses, tissue engineering, wound dressings and drug delivery [7]. A wide variety of synthetic and natural polymers are currently used to prepare hydrogels and novel formulations are continually being developed to satisfy the needs of each particular application. In recent years, thiol-ene systems have been used for the preparation of hydrogels through either photopolymerization or base-catalyzed Michael addition reactions [1-4]. Thiol-ene polymerization reactions occur under ambient conditions and are not inhibited by oxygen and water. These features permit the use of thiol-ene systems in the manufacture of hydrogels for biomedical applications [5]. This study shows the preparation and characterization of hydrogels based on a thiol-methacrylate system.

Recently, nanocomposite materials containing silver nanoparticles have attracted much attention due to their potential applications in optics and electronics. In addition, antimicrobial properties of silver nanoparticles make them very attractive for biomedical applications [8]. The two synthetic routes to prepare silver-polymer nanocomposites are designated as *in-situ* and *ex-situ* methods. The *ex-situ* method involves the dispersion of silver nanoparticles produced previously into a polymer while the *in-situ* method consists in the synthesis of the nanoparticles directly in the polymer matrix [9-10]. A critical aspect in the preparation of nanocomposite materials is the irreversible aggregation of the nanosized particles. Capping molecules which bind onto the surface of the particles are commonly used to avoid particle aggregation. Adsorption of organothiol molecules on silver surfaces is widely used to prevent the formation of particle aggregates. In this research, stable silver nanoparticles were prepared in-*situ* in a tetra functional thiol monomer. The dispersions of nanosized silver particles in thiol were mixed with a bi functional methacrylate monomer and the resultant resins were photoactivated with 2,2-Dimethoxy-2-phenylacetophenone or Camphor-quinone and then photopolymerized by irradiation with UV or visible light respectively.

OUR RESEARCH

This section is divided into three parts concerned first with the polymerization of thiol-methacrylate networks, followed by studies on the synthesis and characterization of hydrogels based on thiol–methacrylate systems. The last section shows results on the preparation of silver nanoparticles in a tetra functional thiol monomer, followed by studies of photopolymerization of a thiol-methacrylate network containing different amounts of silver nanoparticles.

POLYMERIZATION OF THIOL-METHACRYLATE SYSTEMS

Photopolymerization studies were carried out in mixtures of pentaerythritol tetra(3-mercapto propionate) (TMP, >95%, Sigma Aldrich, USA) with Bisphenol A EthoxylateDimethacrylate (EMAS, 98%,Mn~540,

Esstech, Essington, USA) and Bisphenol A ethoxylatedimethacrylate (EMAL, Mn~1700 from Sigma Aldrich, USA). Thiol-methacrylate mixtures were activated for visible light polymerization by the addition of 1 wt. % CQ. The light source used to photocure the resins was assembled from a 5W light-emitting diode (LED) with its irradiance centered at 470 nm (High power LED SML-LXL99USBC-TR/5 from Lumex, USA). UV polymerization was carried out with 2,2-Dimethoxy-2-phenylacetophenone (DMPA, 99%, Sigma Aldrich) using an UV light source assembled from a light-emitting diode (LED) with its irradiance centered at 365 nm (OTLH-0480-UV, Optotech, USA). The structure of the monomers and photo initiators is shown in Scheme 1.

Thiol-ene polymerization can be photoinitiated by initiating species generated by either α-bond cleavage of an excited aryl aliphatic ketone or the excitation of a ketone followed by hydrogen transfer [1-3]. The two-step process typical of the thiol-ene free-radical chain reaction is photoinitiated as illustrated in Scheme 2. The excitation of DMPA under UV-light irradiation ($\lambda = 365$ nm) produces benzoyl and dimethoxybenzyl radicals. A rearrangement of the dimethoxybenzyl radical generates a methyl radical and methyl benzoate. The methyl and benzoyl radicals may add to the double bond or abstract hydrogen from the thiol group. Alternatively, the excitation of CQ under visible light irradiation ($\lambda = 470$ nm), generates the reactive triplet state CQ*, which can react with hydrogen donors to produce pinicol and radicals derived from the hydrogen donor [11-12]. The excitation of CQ followed by hydrogen transfer from the thiol results in a ketyl radical and a sulphur-centered thiyl radical [9, 13-14]. The reaction of thiol and methacrylate monomers to form thiol-methacrylate networks proceeds through a mixed step-chain growth polymerization. The polymerization mechanism involves the homopolymerization of methacrylate monomers through a free-radical chain mechanism in addition to the hydrogen abstraction from the thiol monomers by a chain transfer reaction. Free-radicals derived from thiol groups (thiyl radicals) propagate through the $C = C$ double bond thereby consuming the methacrylate monomer. Thus, when a blend of thiol and methacrylate monomers is polymerized, the radicals resulting from the methacrylate monomer can propagate through

Claudia I. Vallo and Silvana V. Asmussen

another double bond and also participate in chain transfer to thiol. Termination occurs by the coupling of two radical centers.

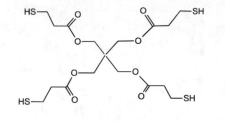

TMP

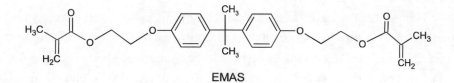

EMAS

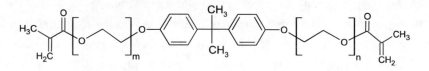

EMAL

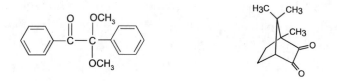

Scheme 1. Structure of the monomers and photo initiators used in this study.

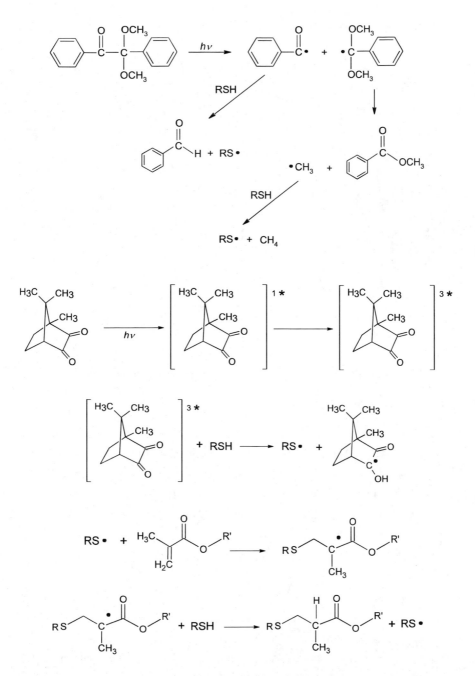

Scheme 2. Photolysis of DMPA and CQ under irradiation nm to produce initiating species for the polymerization of thiol-methacrylate.

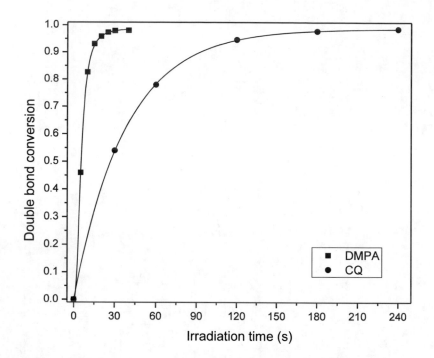

Figure 1. Conversion of methacrylate groups versus irradiation time in EMAL photoactivated with either 1 wt% CQ or 1 wt% DMPA. The specimens were 10 mm in diameter and 3 mm thick.

The conversion of methacrylate groups versus irradiation time was measured at room temperature (20°C) using near infrared spectroscopy (NIR), from the decay of the absorption band located at 6165 cm^{-1}, with a Nicolet 6700 Thermo Scientific. Figure 1 shows the conversion of C = C in TMP-EMAS, at EMAS/TMP molar ratio = 2,photoactivated with DMPA and CQ and irradiated with UV and visible ligh respectively. Plots in Figure 1 show that CQ, in the absence of co-initiator,is an efficient photoinitiator of the thiol-methacrylate systems studied because a fast reaction and high conversions are obtained with concentrations as low as 1 wt% CQ. Moreover, an efficient hydrogen transfer from the TMP to the excited triplet state of CQ, to give a good yield of thiyl radicals is confirmed. The conversion of double bonds was 96% after 120 s of irradiation and it increased up to 100% after 2 h in the dark due to the presence of persistent free radicals trapped in the network. The conversion of the thiol can be assessed from the band located at 2568 cm^{-1} in the mid-IR region, which is associated to the S-H stretch; however thiol absorbance was below detectable limits for the formulations studied. Thus, the conversion of thiol groups was calculated by Raman spectroscopy with an Invia Reflex confocal Raman microprobe (Renishaw). Typical Raman spectra are shown in Figure 2. The conversion of methacrylate C = C double bonds was followed by the decay of the band located at 1642 cm^{-1}, while the conversion of S-H group was assessed by the decay of the band at 2575 cm^{-1}. Details of the FTIR and Raman techniques were reported elsewhere [15].

Table 1. Conversion of thiol (SH) groups at 100% conversion of C = C groups in resins photoactivated with 1 wt.% CQ

	SH conversion	Tg (°C)
Pure EMAL	-	-35
EMAS:TMP r = 2	0.46	38
EMAS:TMP r = 1	0.26	8
EMAL:TMP r = 2	0.48	-37
EMAL:TMP r = 1	0.24	-41

r is the EMAS:TMP or EMAL:TMT molar ratio. The glass transition temperature of the networks (Tg) is also shown

While Raman spectra (Figure 2) showed a complete consumption of the methacrylate groups (see disappearance of the band at 1642 cm^{-1} in Figure 2), a fraction of thiol groups remained unreacted as a result of homopolymerization of the methacrylate monomer. The conversion of thiol groups at 100% conversion of C = C groups is presented in Table 1. The ratio of homopolymerization through "ene" functional groups vs. thiol-ene addition has been reported by previous researchers [16-19]. Depending on the ene monomer used, the reaction progresses with identical consumption of ene and thiol groups, or the ene group shows higher conversion than the thiol group as a result of homopolymerization. Bowman et al. found that for some thiol-ene systems the polymerization is representative of a true step growth polymerization with nearly equivalent consumption of the two functional groups [16-17]. Conversely for stoichiometric thiol-acrylate systems, the conversion of the acrylate functional groups was roughly twice that of the thiol functional groups [16-17]. Similarly, Lecamp et al. [18], showed that in a thiol-methacrylate system the methacrylate homopolymerization was faster than the thiol-ene addition so that the reaction was stopped because of the complete consumption of the methacrylate double bonds.

Visible-light polymerization of methacrylate and epoxy resins is generally photoinitiated by the pair camphorquinone (CQ)/amine [20-22]. However, amines have intrinsic disadvantages such as odor, toxicity, and migration in UV-curing technology. An issue of major concern in packaging systems is the migration of low molecular weight species from the coating material into a packaged product. Results obtained in this research show that a great benefit of using CQ to initiate the photopolymerization of thiol-methacrylate systems is that no amine co-initiator is required. CQ exhibits a strong yellow color due to the presence of an absorbing diketone chromophore in its structure. As a result of the presence of unreacted CQ, the photocured TMP-EMAS and TMP-EMAL networks were yellow in color. The photobleaching of the remaining CQ to give colorless polymers is can be described as follows. Under irradiation of CQ with visible-light the conjugation of the α-dicarbonyl chromophore is destroyed, thus decomposing into colorless products [20]. Therefore, the photobleaching of CQ can be examined by monitoring changes in absorbance during irradiation by UV-vis spectroscopy. Photolysis of CQ in thiol-methacrylate resins was assessed by measuring changes in absorbance at 470 nm during irradiation. Absorption spectra were acquired with an UV-vis spectrophotometer 1601 PC Shimadzu. The resins were contained in cells (3 mm) constructed from two quartz microscope slides separated by a PTFE gasket. An identical cell containing the monomer without CQ was used as the reference. Figure 3 illustrates spectral changes during irradiation of CQ in TMP-EMAS showing a continuous decrease in absorbance during irradiation. 3mm thick samples containing 1 wt. % CQ became colorless after about 9 min irradiation. The rate of photobleaching of CQ depends on the radiation absorbed and the quantum yield and [23]:

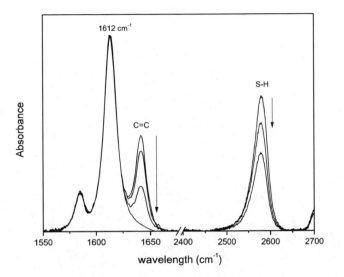

Figure 2. Characteristic peaks in Raman spectra of thiol/methacrylate mixtures. The band at 2575 cm⁻¹ is assigned to the S-H group. The band representing the methacrylate double bond is located at 1642 cm⁻¹. The band at 1612 cm⁻¹, was selected as internal reference band.

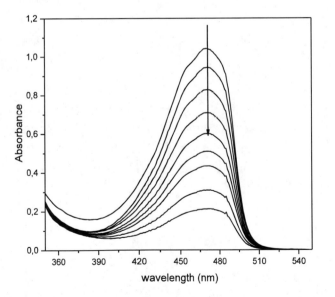

Figure 3. Typical spectral changes showing the photobleaching of CQ in TMP-EMAS resin containing 1 wt % CQ during irradiation at 470 nm. The specimen was 10-mm diameter and 3-mm thick.

Claudia I. Vallo and Silvana V. Asmussen

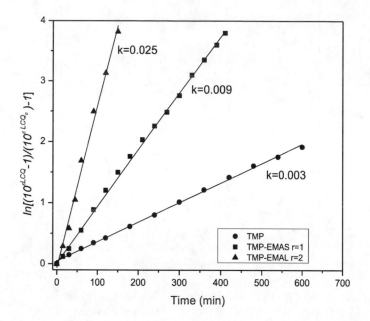

Figure 4. Plots of Eq. (2) for TMP-EMAS, and TMP. Resins containing 1 wt. % CQ were irradiated at 470 nm. The specimen was 10-mm diameter and 3-mm thick. The slopes of the lines (k) are shown for comparison.

$$-\frac{dCQ}{dt} = \frac{\Phi I_{abs}}{L} = \frac{\Phi I_0 (1 - e^{-\varepsilon LCQ})}{L} \qquad (1)$$

where CQ is the concentration of CQ (in mol liter^{-1}, ε is the molar absorption coefficient of CQ, I_0 is the irradiance of the light source (in moles photons s^{-1} cm^{-2}), L is the thickness of the sample, and Φ is the quantum yield of the photoinitiator consumption, i.e., the fraction of photoinitiator reduced per absorbed photon. Integrating Eq. (1) yields:

$$ln\left[\frac{10^{\varepsilon LCQ} - 1}{10^{\varepsilon LCQ_0} - 1} \right] = -\Phi \varepsilon I_0 t \qquad (2)$$

where the product ($\Phi \varepsilon I_0$) is the rate constant for the photo-decomposition of CQ and CQ_0 is the initial concentration of CQ. Figure 4 Illustrates typical plots of Eq. (2). Experimental measurements of absorbance vs. irradiation time resulted in satisfactory fits to a first-order kinetics for the photobleaching of CQ in all the thiol-methacrylate resins studied. This trend is analogous to that observed during the photodecomposition of CQ in methacrylate resins photoactivated with CQ/amine [20]. Photobleaching of CQ in thiol-methacrylate resins occurs because of the presence of labile hydrogen atoms in TMP, EMAS and EMAL. In the photolysis of CQ hydrogen abstraction from thiol groups in TMP and oxyethylene units in EMAS and EMAL structures are competitively involved. Therefore, the different rate constant values shown in Figure 4 are attributed to the presence of different sources of abstractable hydrogen atoms. It is worth noting here that the use of photobleaching initiators is strategic for the cure of thick sections because the consumption of initiator results in an increase in light intensity along the radiation path [20, 24]. Results from UV-vis studies in Figure 3 demonstrate that the use of CQ permits the photopolymerization of the thiol-methacrylate networks in thick sections.

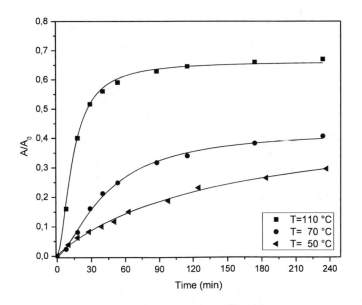

Figure 5. Plots of absorbance at 470 nm versus time at different temperatures. 3 mm thick samples of EMAS:TMP at molar ratio equal to 2 were photobleached at room temperature and then placed in the controlled temperature accessory of the UV-vis spectrophotometer.

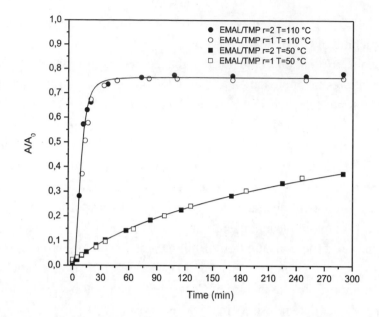

Figure 6. Plots of absorbance at 470 nm versus time at different temperatures. 3 mm thick samples of EMAL:TMP were photobleached at room temperature and then placed in the controlled temperature accessory of the UV-vis spectrophotometer. No differences are observed between samples having r = 1 or r = 2.

Previous studies have shown that some diary ketones used to photo initiate thiol-ene systems remain unconsumed at the end of the polymerization and jeopardize the long-term color stability (yellowing) of the cured films [1]. Yellowing in cured films and bulk materials is an adverse effect caused by the presence of photoproducts from initiator residues [25-26]. The color stability of thiol-methacrylate networks photoactivated with CQ was examined by monitoring changes in the absorbance at 470 nm resulting from the regeneration of CQ by means of UV-vis spectroscopy. Films based on thiol-methacrylate systems were totally photobleached at room temperature and then were immersed in the controlled temperature accessory of the UV-vis spectrophotometer in order to monitor the change in absorbance at 470 nm as a function of time. Figure 5 shows the increase in absorbance of CQ under heating of a thiol-methacrylate polymer having a molar ratio EMAS/TMP equal to 2. Figure 6 shows the progress of the absorbance of CQ in EMAL/TMP at molar ratios EMAL/TMP equal to 2 and 1. As shown in Figures 5-6 the rate of change in absorbance at 470 nm is strongly influenced by the temperature of the medium. The absorbance at 470 nm in 3 mm thick specimens (not shown in Figures 5-6) increased slowly to 0.2 after 4 weeks at $20 \pm 2°C$. The changes in color during heating of thiol-methacrylate networks depicted in Figures 5-6 are attributed to reactions of hydrogen atom transfer of thiols with ketones. Thiyl radicals participate in hydrogen atom transfer reactions which convert the ketyl radical intermediates back to starting materials, in competition with ketyl radicals coupling reactions to form pinacol [27].Cohen et al. studied reactions occurring between thiols and excited triplet benzophenones in 2-propanol and acetonitrile [27].Scheme 2 illustrates that, in analogy with Benzophenone, the thiyl radical can disproportionate with the ketyl radical to regenerate CQ and thiol. Results presented in Figs 5-6 demonstrate that, as a result of reactions leading to recombination of CQ, the long term stability of films and bulk materials based on thiol-ene systems is jeopardized.

In addition to the thiol-methacrylate free radical reaction described before (Scheme 3-a), thiol-methacrylate addition reactions can proceed through base- or nucleophile-catalyzed Michael addition reaction. [5-6]. In the presence of primary/secondary amines; the sulfur atom is deprotonated which results in the formation of ammonium cation and thiolate anion (see Scheme 3-b). The thiolate anion produced adds into the C = C double bond at the β-carbon generating an intermediate carbon-centered anion which is a very strong base. When the enolate anion picks up a proton either from the ammonium cation or from a thiol group the final thiol- methacrylate product is formed. The Michael addition between EMAS/TMP and EMAL/TMP in the presence of Propylamine (PAm, >99%, Sigma Aldrich) as catalyst was carried out at ambient temperature. The conversion of C = C and SH as a function of time was studied in mixtures having methacrylate-to-thiol molar ratio of either 1 or 2. Figure 7 shows the conversion of C = C bonds (in EMAS/TMP and EMAL/TMP at molar ratios methacrylate/thiol = 2:1 by NIR spectroscopy calculated from the decay of the absorption band located at 6165 cm^{-1}.

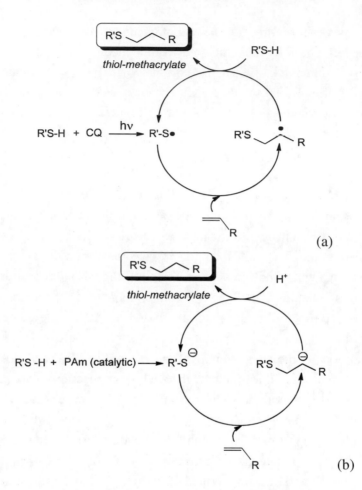

Scheme 3. (a) Idealized thiol-ene polymerization by free-radical addition reaction. (b) Idealized thiol-ene polymerization by base-catalysed Michael addition reaction.

The higher polymerization rate in EMAL/TMP mixtures compared to that in EMAS/TMP mixtures in Figure 7 is attributed to a higher PAm/TMP molar ratio in resins prepared with the higher molar mass methacrylate. The effect of the methacrylate/thiol molar ratio on the polymerization rate is shown in Figure 8. The faster polymerization reaction in mixtures having r= 1 in Figure 7 is attributed to a faster generation of thiolate anions by deprotonation of the thiol groups in mixtures containing thiol in excess. EMAS/TMP and EMAL/TMP specimens used for measurements of C = C conversion by NIR were examined by Raman spectroscopy. Raman spectra showed a complete conversion of SH groups for the networks having a methacrylate/thiol molar ratio equal to 2:1. Conversely, 50% of the SH groups remained unreacted in mixtures with thiol in excess (methacrylate/thiol molar ratio = 1).The conversion of thiol groups at 100% conversion of C = C groups is shown in Table 2.

Table 2. Conversion of thiol (SH) groups at 100% conversion of C = C groups in resins containing 2 wt.% PAm

	SH conversion	T_g (°C)
EMAS:TMP r = 2	1	10
EMAS:TMP r = 1	0.5	-3.5
EMAL:TMP r = 2	1	-37
EMAL:TMP r = 1	0.5	-37

r is the EMAS:TMP or EMAL:TMT molar ratio. The glass transition temperature of the networks (T_g) is also shown

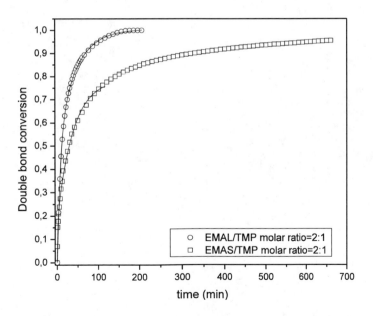

Figure 7. Conversion of C = C versus time in mixtures prepared with EMAS-TMP and EMAL-TMP containing 1 wt% PAm. The conversion of C = C groups was calculated using NIR spectroscopy from the decay of the absorption band located at 6165 cm^{-1}.

In spite of the large proportion of unreacted thiol groups the percentage of extractable thiol monomers is very low because the TMP monomer is tetra functional and remains as pendant chains in the final networks [28]. Thiol-ene networks are usually prepared from stoichiometric proportions of reactants. Recently, Khutoryanskiy and co-workers reported that materials with mucoadhesive properties, i.e., able to stick on mucosal surfaces, can be prepared from non-stoichiometric thiol-ene formulations [29]. Studies of thiol–methacrylate reactions with non-stoichiometric ratio of reagents for the synthesis of hydrogels containing unreacted thiol functional groups were encouraged by the results reported in the aforementioned reports.

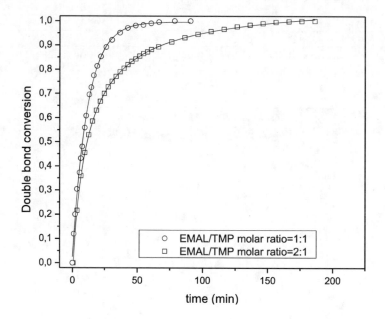

Figure 8. Conversion of C = C versus time in mixtures prepared with different EMAL/TMP molar ratios, containing 2 wt% PAm. The conversion of C = C groups was calculated using NIR spectroscopy from the decay of the absorption band located at 6165 cm^{-1}.

No specific study of adhesion was made in the present study; however, it was observed that there was a marked tendency of hydrogels with thiol in excess to adhere to a variety of substrates, including metals, glass, paper and skin. This feature makes them especially attractive in applications for which adhesion is particularly critical [30]. The glass transition temperature (Tg) is a very important property of a polymer network because it normally defines the type of applications a polymer is suitable for. Consequently, the Tg of the prepared thiol-methacrylate networks was assessed by dynamic mechanical thermal analysis (DMTA). DMTA measurements were performed in torsion deformation mode from -70 to 100°C at a heating rate of 5°Cmin^{-1} using an MCR 301 rheometer (Anton Paar GmbH). Rectangular specimens were tested at 1 Hz. Previously, a strain sweep test of a representative sample was performed at 20°C and 1 Hz in order to determine the linear viscoelastic range and select a strain value to apply in temperature sweeps.

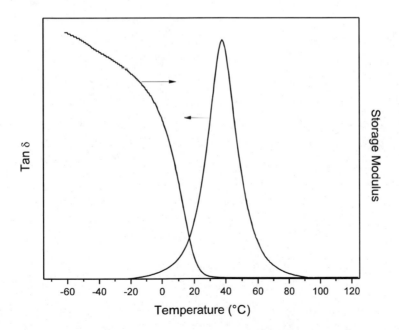

Figure 9. Tanδ and storage modulus versus temperature for TMP:EMAS 1:2 mol ratio.

The selected strain value was 0.05%. A small axial force (~ -0.5 N) was applied in all test specimens in order to maintain a net tension. Details concerning this technique have been reported elsewhere [15]. Representative plots of storage modulus and tanδ versus temperature obtained from DMTA analysis are shown in Figure 9 while the Tg values are presented in Tables 1-2.Thiol-ene networks are distinguished by a high network uniformity which is characterized by the width of the glass transition region [31]. This is commonly quantified by the full width at half-maximum for the tanδ in the transition region. The thiol-methacrylate polymer based in EMAS-TMP displays a half-height widths Tg equal to 18°C. Conversely, photocured multifunctional methacrylates exhibit a much broader glass transition extending over about 140°C [2]. Tables 1-2 show that the multifunctional TMP in combination with EMAS and EMAL produces polymeric networks with a wide variety of Tg values. Both hard glassy polymers and flexible elastomers are produced by varying the structure of the methacrylate monomer and the methacrylate/thiol molar ratio. The Tg is governed by the crosslink density, the average molecular weight between crosslinks, or the concentration of elastic chains. The Tg of EMAS/TMP systems is affected by the EMAS/TMP molar ratio (r). The lower Tg value in the EMAL/TMP network with thiol in excess is attributed to a lower concentration of elastic chains and the related lower crosslink density [32]. Similarly, the lower values of Tg in EMAL/TMP polymers is explained in terms of a inferior crosslinking density because of both the higher molar mass of EMAL and the intrinsic flexibility of the chemical bonds present EMAL. Even though the principal structural factor affecting the Tg of rigid networks is the concentration of elastic chains, its effect on flexible networks is less important [32]. This is possibly so because the relaxation of flexible networks preferentially occurs through "hinges" present in their chemical structures (i.e., the ether bonds of EMAL), being much less affected by the concentration of crosslinking points. Consequently, the effect of the EMAL/TMP molar ratio on Tg is much less important for these networks.

HYDROGELS BASED ON
THIOL-METHACRYLATE NETWORKS

Hydrogels from EMAL copolymerized with the TMP were synthesized through free-radical addition reaction in resins photoactivated with 1wt% CQ and irradiated with visible light at 465 nm.EMAL was also copolymerized with TMP by Michael addition reaction catalyzed with 1wt% PAH (Scheme 4). Hydrogels from EMAL-TMP mixtures were prepared at either EMAL-to-TMP molar ratio of 2:1 (stoichiometric proportion) or EMAL-to-TMP molar ratio of 1:1 (thiol in excess).

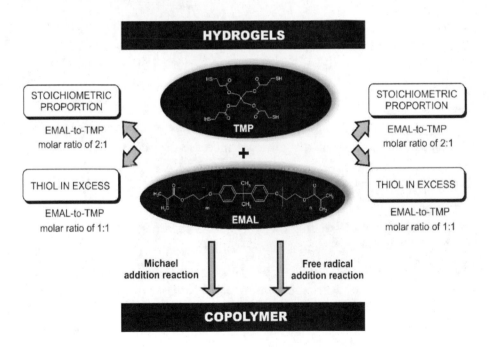

Scheme 4. Methods used for preparation of hydrogels derived from TMP-EMAL.

Since the EMAL and TMP monomers and their mixtures are liquid they can be placed in a mold to produce specimens of the desired shape. The liquid resins containing the photo initiator system were sandwiched between two thin glass plates separated by a 3mm thick rubber sheet with either a circular or a rectangular hole. The hole of the rubber sheet was filled with the reactive mixture and the assembly was held using small clamps. The resins were polymerized at room temperature to form 1mm thick rectangular slabs (10mm× 25 mm).

Water uptake in hydrogels was measured using a standard gravimetric sorption technique. Slabs or discs of hydrogels were first weighed and then immersed in distilled water at $25 \pm 1°C$. Swollen specimens were removed from solutions at regular time intervals, blotted with filter paper and weighed. The degree of swelling (Sw) was calculated as follows:

$$S_w(\%) = \frac{m_t - m_d}{md} \times 100 \tag{3}$$

Where m_t and m_d are the weights of swollen and dry specimens, respectively. Figure 10 shows the water uptake of EMAL-TMP networks prepared by free radical and Michael addition reactions. The water sorption of the EMAL-based hydrogel is also shown for comparison. No significant differences in water uptake are observed between specimens prepared from EMAL–TMP photo initiated by CQ and photo cured specimens prepared from pure EMAL. On the other hand, a significant increase in water uptake is seen in the EMAL-TMP hydrogels prepared by the PAm- catalyzed Michael addition reaction. The swelling of hydrogels is produced by the presence of hydrophilic groups in the polymer structure. In addition, a critical factor that governs the degree of swelling of hydrogels is the crosslink density of the network [33]. Hydrogels with a low crosslinking density have a more open structure and will swell more than highly crosslinked hydrogels. As illustrated in Scheme 3, each thiol group reacts with one methacrylate double bond, while in a chain-growth radical polymerization each double bond is coupled to two monomers. Thus, the thiol Michael addition reaction results in a comparatively lower crosslinking density than the free radical addition reaction because the effective monomer functionality in the step-growth network is half that observed in chain-growth radical polymerizations. Trends of swelling shown in Figure 10 are attributed to a balance between the crosslink density and the mass fraction of hydrophilic groups of the network.

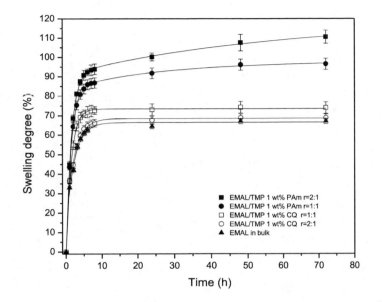

Figure 10. Degree of swelling of EMAL–TMP networks prepared through free radical and amine-catalysed Michael addition reactions at various EMAL/TMP molar ratios. The swelling of EMAL photopolymerized in bulk is shown for comparison.

The Tg values of the hydrogels prepared from EMAL are shown in Tables 1-2 while typical plots of tanδ versus temperature from DMTA studies are presented in Figure 11. EMAL monomer forms a moderately crosslinked polymer because of its high molar mass and contains a high proportion of flexible ether groups; therefore, the EMAL-derived network displays a low Tg (Table 1). Similarly, EMAL-TMP polymers formed by thiol Michael addition reactions contain a high proportion of flexible thioether bonds. The identical values of Tg of networks synthesized from different EMAL-to-TMP molar ratios contrast with many experimental results reported in the literature showing the dependence of Tg on parameters such as crosslink density and average molar mass between crosslinks [32]. Although in rigid networks the most important structural factor affecting Tg is the average molar mass between crosslinks, its effect on flexible networks is much less important. The relaxation of flexible networks preferentially occurs through 'hinges', such as ether and thioether bonds present in their structures, and is much less affected by the concentration of crosslinking points [32].

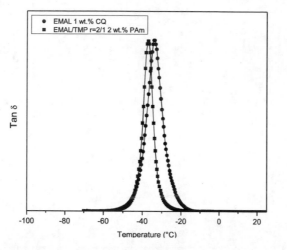

Figure 11. Typical plot of tan δ versus temperature for thiol–methacrylate prepared by PAm-catalysed Michael addition reaction. Parameter r is the EMAL-to-TMP mole ratio.

In analogy with poly(ethylene glycol) dimethacrylate, which shows outstanding biocompatibility, [34] EMAL is expected to be suitable for biomedical applications. The structure of EMAL resembles that of methacrylate monomers that have been successfully used in dental restorative composites during the last two decades [35]. Similarly, the use of TMP for the preparation of biomaterials for drug delivery applications and tissue engineering has been reported [28]. CQ has demonstrated to be a biocompatible photo initiator after three decades of use in light-cured dental resins. All these factors make EMAL-TMP attractive for the preparation of hydrogel for biomedical applications.

THIOL-METHACRYLATE NANOCOMPOSITES CONTAINING SILVER NANOPARTICLES

Colloidal dispersions of silver particles in TMP were prepared by *in situ* reduction of silver nitrate ($AgNO_3$, 99%, Sigma Aldrich USA) with 2,6-di-tert-butyl-p-cresol (BHT, > 99%, Sigma Aldrich USA). First, $AgNO_3$ was dissolved in absolute ethanol. Then, the solution of $AgNO_3$ in ethanol was mixed with different amounts of TMP in order to prepare dispersions with different amounts of silver in the range 200-2500 ppm (Scheme 5). After complete evaporation of the ethanol, the prepared colloidal dispersions of silver particles in TMP were mixed with EMAS at molar ratio EMAS to TMP = 1. Details of the procedure were reported elsewhere [36].

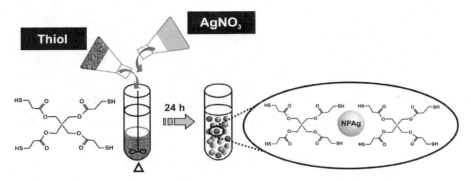

Scheme 5. Preparation of colloidal dispersions of Ag nanoparticles in TMP by *in situ* reduction of silver nitrate with BHT.

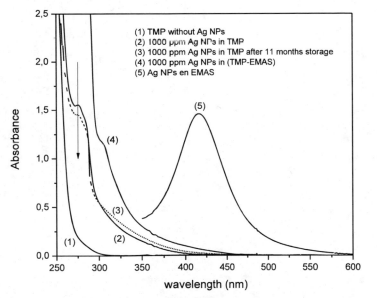

Figure 12. UV–vis spectra of dispersions of Ag NPs in TMP, EMAS from Ref. [10], and a mixture TMP-EMAS 1:2 mol ratio.

Figure 12 shows the UV-vis absorption spectra of a dispersion of nanosized silver particles in TMP. The absorption spectra of Ag NPs (~10 nm in diameter) prepared in the EMAS monomer is also shown for comparison [10]. The UV-vis absorption spectrum of a dispersion of silver particles in a mixture EMAS-TMP 2:1 mol ratio shown in Figure 12 is broader than that of the dispersion of Ag in TMP. According to Mie's theory, the surface plasmon resonance absorption (SPR) position and shape are closely related to the particle size and shape [37]. The SPR of the Ag NPs in the mixture TMP-EMAS starts in the visible region (λ = 450 nm) indicating a change in the size of the nanoparticles. This difference in size of Ag NPs prepared in different monomers is attributed to a stronger interaction of TMP with the silver particles compared with EMAS. TEM microscopy studies were carried out in order to examine the morphology of the particles in the TMP-EMAS network. Details of the technique and sample preparation were reported elsewhere [36]. TEM images in Figure 13 show that the silver particles are distributed in the polymer and also form aggregates. The formation of aggregates is the result of a reduction in the TMP/Ag mass ratio in the preparation of dispersions of Ag in TMP-EMAS by adding EMAS to dispersions of Ag nanoparticles in TMP. The decrease in the amount of thiol results in a disruption of the protection layer and, therefore, agglomeration of the particles.

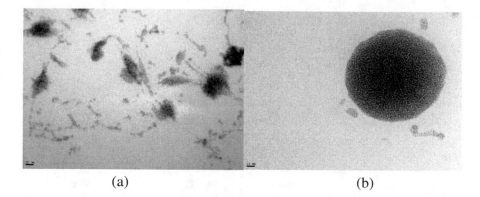

(a)   (b)

Figure 13. TEM images of 500 ppm Ag NPs dispersed in TMP-EMAS. (a) The length of the bar is equivalent to 20 nm. (b) TEM pictures showing agglomerations of Ag NPs in a dispersion of 2500 ppm Ag NPs in TMP-EMAS.

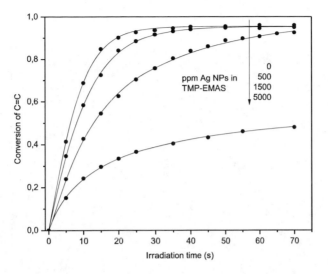

Figure 14. Conversion of methacrylate groups in TMP-EMAS containing different amounts of Ag NPs (ppm). Samples containing 1 wt% DMPA were irradiated at 365 nm.

Mixtures thiol-methacrylate containing different proportions of nanosized silver particles were photoactivated with either 1 wt. % of DMPA or 1 wt. % of CQ. Resins containing CQ were photo polymerized with a 470-nm LED while resins containing DMPA were photo polymerized with a 365-nm LED. The conversion of C = C were calculated by NIR spectroscopy from the decay of the absorption band located at 6165 cm^{-1}. The conversion of C = C in TMP-EMAS mixtures containing different amounts of Ag nanoparticles, photoactivated with DMPA and CQ is shown in Figures 14 and 15 respectively. It is seen that the cure of the resins became more difficult as the Ag nanoparticles concentration increased. The final conversions in resins containing 0 or 1500 ppm Ag NPs were similar, however, the polymerization rate decreased appreciably on increasing the amount of silver particles. When an assembly of nanoparticles is irradiated by a plane wave, the incident light is scattered and absorbed by each particle to a certain extent [37]. This is attributed to the fact that the transmitted light decreases along the propagation direction of the incident light because the light is scattered and absorbed by the particles.

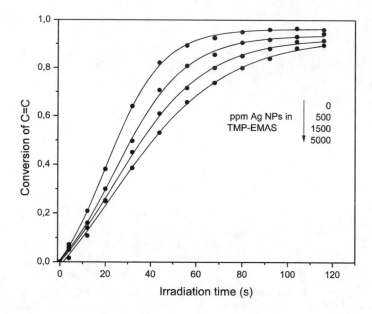

Figure 15. Conversion of methacrylate groups in TMP-EMAS containing different amounts of Ag NPs (ppm). Samples containing 1 wt% CQ were irradiated at 470 nm.

Figure 16 shows the UV-vis spectra of the photo initiators DMPA and CQ, a dispersion of Ag NPs in TMP-EMAS, and the wavelengths of the emitted light from the UV and visible light sources. The UV source emits in a narrow wavelength range centered at 365 nm while the Ag particles absorb light in the range 350–550 nm. Consequently, a fraction of the incident light is absorbed by the particles. The presence of light absorbing species in the resin results in attenuation of the light intensity. The reduced polymerization rate with increasing silver content is attributed to the related losses which increase with increasing amounts of particles [38]. The reduction in the polymerization rate in resins photo activated with CQ is much less marked (Figure 15). This is explained in terms of a lower overlap of the spectral emission of the visible light source with the SPR of the Ag NPs (Figure 16). From these results, it emerges that in order to cure layers of resins containing high proportions of nano sized silver particles; the light source-photo initiator pair must be carefully selected.

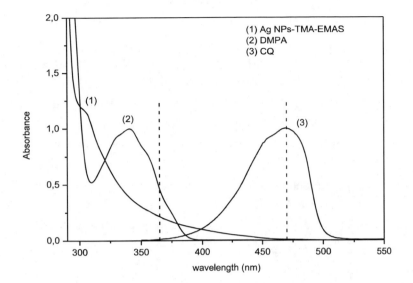

Figure 16. UV–vis spectra of DMPA and CQ, and resin TMP-EMAS containing Ag NPs. The dashed lines represent the wavelengths of the emitted light from the UV and LEDs.

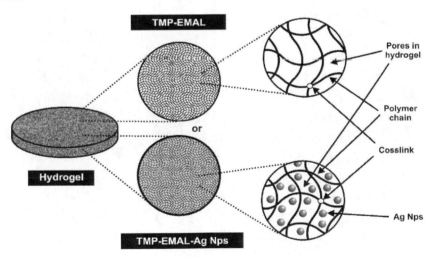

Scheme 6. Structure of hydrogels prepared from colloidal dispersions of Ag nanoparticles in a TMP-EMAL mixture.

From results presented in this report it emerges that thiol- methacrylate resins based on MP-EMAS are efficiently polymerized with visible light by CQ in the absence of amine and thereby prevents the complications related to the use of amines. In addition, silver and silver ions have long been known to exhibit powerful antimicrobial activity [39-41]. We found that silver nanoparticles can be synthesized by reduction of $AgNO_3$ with BHT which is a low toxicity reducer. In fact, BHT is an antioxidant that is often added to foods to preserve fats and is present in formulations of light-cured dental resins to prevent premature polymerization during storage. Thus, BHT is an attractive reducing agent for applications that demand low toxicity. This makes the light-cured TMP-EMAL containing Ag NPs very attractive for the preparation of biomaterials and coating with improved biocompatibility. In particular, as illustrated in Scheme 6, the nanocomposites can be used for the production of hydrogels based on thiol-methacrylate networks.

CONCLUSION

Thiol-methacrylate mixtures photo activated with DMPA and CQ are efficiently polymerized by UV and visible light respectively. Moreover, the photo polymerization with CQ progresses without the addition of co initiator amine. Unfortunately, the presence of reactions leading to recombination of CQ jeopardizes the long-term durability of films and bulk materials based on thiol-ene systems. Thiol-methacrylate films exhibit yellowing over time because of regeneration of CQ through hydrogen transfer reactions between ketyl radicals and thiyl radicals. If color stability is a factor to take into account, the photo polymerization must be carried out with colorless photo initiators such as DMPA.

The EMAL methacrylate monomer results very attractive for the preparation of hydrogels following different synthesis pathways. Hydrogels formulated from EMAL-TMP, by PAm-catalyzed Michael addition reaction, show a marked increase of water sorption compared with the hydrogels derived from pure EMAL. On the other hand, no significant differences in water sorption are observed between hydrogels formulated from EMAL and EMAL-TMP prepared by photo polymerization. These trends are explained in terms of a balance between the crosslinking density of the networks and the proportion of hydrophilic groups.

Stable dispersions of silver nanoparticles in TMP are easily prepared by *in situ* reduction of $AgNO_3$ with BHT. Thiol-methacrylate mixtures containing silver nanoparticles can be photo cured provided the emission range of the light source does not overlap with the surface plasmon resonance absorption band of the nanoparticles. The reactants needed for the synthesis have low toxicity and silver nanoparticles show antimicrobial activity. Thus, the nanocomposite materials synthesized in this study are very attractive for the preparation of biomaterials and coating with improved biocompatibility.

REFERENCES

[1]   Hoyle CE, Lee TY, Roper T, Thiol-enes: chemistry of the past with promise for the future, *J. Polym. Sci. Part A Polym. Chem.* 42 (2004) 5301.

[2]   Hoyle CE, Bowman CN, Thiol-ene click chemistry, *Angew. Chem. Int. Ed.* 49 (2010) 1540.

[3]   Tasdelen MA, Kiskan B, Yagci Y, Externally stimulated click reactions for macromolecular syntheses, *Prog. Polym. Sci.* 52 (2016) 19.

[4]   Lowe AB, Thiol-ene"click" reactions and recent applications in polymer and materials synthesis, *Polym. Chem.* 1 (2010) 17.

[5]   Aimetti AA, Machen AJ and Anseth KS, Poly(ethylene glycol) hydrogels formed by thiol-enephoto polymerization for enzyme-responsive protein delivery, *Biomaterials* 30 (2019) 6048.

[6]   Hoyle CE, Lowe AB and Bowman, Thiol-click chemistry: a multifaceted toolbox for small molecule and polymer synthesis, *Chem Soc Rev* 39 (2010) 1355.

[7]   Sirousazar M, Forough M, Farhadi K, Shaabani Y and Molaei R, Hydrogels: properties, preparation, characterization and biomedical applications in tissue engineering, drug, delivery and wound care, in *Advanced Healthcare Materials,* ed. by Tiwari A. Wiley, Hoboken, NJ, ch. 9 (2014).

[8]   Pozo Perez D (Ed.), *Silver Nanoparticles*, InTech, 2010 (ISBN: 978-953-307-028-5).

[9]   Kassaee MZ, Mohammadkhani M, Akhavan A, Mohammadi R, In situ formation of silver nanoparticles in PMMA via reduction of silver ions by butylated hydroxytoluene, *Struct. Chem.* 22 (2011) 11.

[10] Asmussen SV, Vallo CI, Synthesis of silver nanoparticles in surfactant-free light-cured methacrylate resins, *Colloids Surf. A: Physicochem. Eng. Aspects* 466 (2015) 115.

[11] Uygun M, Tasdelen MA, Yagci Y, Influence of type of initiation on thiol-ene "click" chemistry, *Macromol. Chem. Phys.* 211 (2010) 103.

[12] Xiao P, Dumur F, Frigoli M, Graff B, Morlet-Savary F, Wantz G, Bock H, Fouassier JP, Gigmes D, Lalevée J, Perylene derivatives as photo initiators in blue light sensitive cationic or radical curable films and panchromatic thiol-enepolymerizable films, *Eur. Polym. J.* 53 (2014) 215.

[13] Battocchio C, Meneghini C,Fratoddi I, Venditti I, Russo MV, Aquilanti G, Maurizio C, Bondino F, Matassa R, Rossi M, Mobilio S, Polzonetti G, Silver nanoparticles stabilized with thiols: a close look at the local chemistry and chemical structure, *J. Phys. Chem.* C 116 (2012) 19571.

[14] Ansar SM, Perera GS, Gomez P, Salomon G, Vasquez ES, Chu IW, Zou S, Pittman CU, Walters KB, Zhang D, Mechanistic study of continuous reactive aromatic organothiol adsorption onto silver nanoparticles, *J. Phys. Chem.* C 117 (2013) 27146.

[15] Asmussen S, Schroeder W, dell'Erba I and Vallo CI, Monitoring of visible light photopolymerization of an epoxy/dimethacrylate hybrid system by Raman and near-infrared spectroscopies, *Polym Test* 32 (2013) 1283.

[16] Cramer NB, Bowman CN, Kinetics of thioleene and thioleacrylate photopolymerizations with real-time Fourier transform infrared, *J. Polym. Sci. Part A Polym. Chem. 39* (2001) 3311.

[17] Cramer NB, Reddy SK, O'Brien AK, Bowman CN, Thiol-ene photo polymerization mechanism and rate limiting step changes for various vinyl functional group chemistries, *Macromolecules* 36 (2003) 7964.

[18] Lecamp L, Houllier F, Youssef B,Bunel C, Photoinitiated cross-linking of athiol-methacrylate system, *Polymer* 42 (2001) 2727.

[19] Lee TY, Roper TM, Jonsson ES, Guymon CA, Hoyle CE, Thiol-Enephoto polymerization kinetics of vinyl acrylate/multifunctional thiol mixtures, *Macromolecules* 37 (2004) 3606.

[20] Asmussen SV, Arenas G, Cook WD, Vallo CI, Photoinitiation rate profiles during polymerization of a dimethacrylate-based resin photoinitiated with camphorquinone/amine influence of initiator photobleaching rate, *Eur. Polym. J.* 45 (2009) 515.

[21] Fouassier JP, Lalevée J, *Photo initiators for Polymer Synthesis-scope, Reactivity, and Efficiency,* Wiley-VCH Verlag GmbH &Co. KGaA, Weinheim, 2012.

[22]   Schroeder WF, Asmussen SV, Sangermano M, Vallo CI, Visible light polymerization of epoxy monomers using an iodonium salt with camphorquinone/ethyl-4-dimethyl aminobenzoate, *Polym. Int.* 24 (2013) 430.

[23] Odian G (Ed.), *Principles of Polymerization,* Wiley, NY, 1991, pp. p.222.

[24] Miller G, Gou I, Narayanan V, Scranton A, Modeling of photobleaching for the photoinitiation of thick polymerization systems, *J. Polym. Sci. Part A Polym. Chem.* 40 (2002) 793.

[25] Müller M, Militz H, Krause A, Thermal degradation of ethanolamine treated poly(vinyl chloride)/wood flour composites, *Polym. Degrad. Stabil.* 97 (2012) 166.

[26] Pastorelli G, Cucci C, Garcia O, Piantanida G, Elnaggar A, Cassar M, Strlič M, Environmentally induced colour change during natural degradation of selected polymers, *Polym. Degrad. Stabil.* 107 (2014) 198.

[27] Stone P0, Cohen SG, Catalysis by aliphatic thiol in photoreduction of benzophenone by amines and alcohols, *J. Phys. Chem.* 85 (1981) 1719.

[28] Rydholm AE, Bowman CN and Anseth KS, Degradable thiol-acrylate photopolymers: polymerization and degradation behavior of an *in situ* forming biomaterial, *Biomaterials* 26 (2005) 4495.

[29] Storha A, Mun EA and Khutoryanskiy VV, Synthesis of thiolated and acrylated nanoparticles using thiol-ene click chemistry: towards novel mucoadhesive materials for drug delivery, *RSC Adv* 3 (2013) 12275.

[30] Szilagyi BA, Gyarmati B, Horvat G, Laki A, Budai-Szücs M, Csányi E et al., The effect of thiol content on the gelation and mucoadhesion of thiolated poly(aspartic acid), *Polym Int* 66 (2017) 1538.

[31] McNair OD, Sparks BJ, Janisset AP, Bren DP, Patton DL, Savin DA, Highly tunable thio-ene networks via dual thiol addition, *Macromolecules* 46 (2013) 5614.

[32] Vallo, CI, Frontini, PM, Williams, RJJ, The glass transition temperature ofnonstoichiometric epoxy-amine networks, *J. Polym. Sci. Part B Polym. Phys.* 29(1991) 1503.

[33] Kopĕcek J and Yang J, Hydrogels as smart biomaterials, *Polym Int* 56 (2007)1078.

[34] Di Ramio JA, Kisaalita WS,Majetich GF and Shimkus JM, Poly(ethylene glycol) methacrylate/dimethacrylate hydrogels for controlled release of hydrophobic drugs, *Biotechnol Prog* 21 (2005)1281.

[35] Asmussen SV, Arenas G, Cook WD and Vallo CI, Photobleaching of camphorquinone during polymerization of dimethacrylate-based resins, *Dent Mater* 25 (2009) 1603.

[36] Asmussen SV, Vallo CI, Facile preparation of silver-based nanocomposites via thiol-methacrylate 'click'photopolymerization, *Eur. Polym. J.* 79 (2016) 173.

[37] Kelly KL, Coronado E, Zhao LL, Schatz G, The optical properties of metal nanoparticles: the influence of size, shape, and dielectric environment, *J. Phys. Chem.* B 107 (2003) 668.

[38] Quinten M, *Optical Properties of Nanoparticle Systems,* Wiley-VCH Verlag & Co. KGaA, Germany, 2011.

[39] Colucci G, Celasco E, Mollea C, Bosco F, Conzatti L, Sangermano M, Hybrid coatings containing silver nanoparticles generated *in situ* in a thiol-enephotocurable system, *Macromol. Mater. Eng.* 296 (2011) 921.

[40] Kumar-Krishnan S, Prokhorov E, Hernández-Iturriaga M, Mota-Morales JD, Vázquez-Lepe M, Kovalenko Y, Sanchez IC, Luna-Bárcenas G, Chitosan/silver nanocomposites: synergistic antibacterial action of silver nanoparticles and silver ions, *Eur. Polym. J.* 67 (2015) 242.

[41] Wong KK, Liu X, Silver nanoparticles - the real "silver bullet" in clinical medicine?, *Med Chem. Commun.* 1 (2010) 125.

In: Thiols: Structure, Properties and Reactions   ISBN: 978-1-53615-599-0
Editor: Carlos C. McAlpine   © 2019 Nova Science Publishers, Inc.

Chapter 3

THIONYL CHLORIDE (SOCL$_2$)-MEDIATED HOMOCOUPLING OF THIOLS INTO DISULFIDES AT AMBIENT CONDITIONS

Palani Natarajan[*], Renu Chaudhary, Jaskamaljot Kaur and Paloth Venugopalan

Department of Chemistry and Centre for Advanced Studies
in Chemistry, Panjab University, Chandigarh, India

ABSTRACT

A novel and efficient protocol has been developed for the synthesis of symmetrical disulfides from thiols using thionyl chloride (SOCl$_2$) as a sole oxidizing agent at ambient conditions. Also, a tentative mechanism has been reported for the reaction. Compared to the already reported methods for the synthesis of disulfides, this method has several advantages, including short reaction time, no laborious work-up procedures (by-products are gases and escape from the reaction vessel promptly), a broad

[*] Corresponding Author's E-mail: pnataraj@pu.ac.in.

functional group tolerance, and takes place in solution as well as in solid states.

INTRODUCTION

Aryl disulfide, [1] alkyl disulfide [2] and their derivatives are an important class of organic compounds present in many natural products (oxytocin) and pharmaceuticals (disulfirams). Moreover, disulfide bonds formed between the thiol groups of cysteine residues by the process of oxidation play an important role in the proteins folding [3]. A large number of protocols have been reported for the preparation of disulfides [4]. Among, the most frequently used methodology is the direct oxidation of thiols into disulfides using various reagent systems such as permanganate, [5] manganese dioxide, [6] 2,6-dicarboxypyridinium chlorochromate, [7] cetyl-trimethylammonium dichromate, [8] Fe(III)/sodium iodate, [9] VO(acac)$_2$, [10] molybdate sulfuric acid, [11] tungstophosphoric acid/ NaBrO$_3$, [12] caesium fluoride-celite, [13] bismuth (III) nitrate, [14] Co(II) phthalocyanines, [15] etc., [15] and non-metallic reagents include benzyl-triphenylphosphoniumperoxymonosulfate, [16] trichlo-roiso-cyanauric acid, [17] trichloronitromethane, [18] hydrogen peroxide, [19] bromine,[20] sulfuryl chloride,[21] NO$_2$ gas, [22] activated carbon [23] and DDQ [24].

Despite thionyl chloride (SOCl$_2$) is a toxic reagent, it has been produced more than 60,000 tons per year and largely being consumed for the production of organochlorine compounds in industries and academia [25]. SOCl$_2$ provides only gaseous byproducts, i.e., HCl and SO$_2$ and decomposes readily by hydrolysis [26]. During the synthesis of acid chloride of cysteine, we realized that in the presence SOCl$_2$, cysteine (**1o**) undergoes dehydrogenative self-coupling (Scheme 1) to afford 3,3'-disulfanediylbis (2-aminopropanoic acid) (**2o**, Scheme 1) in a satisfactory yield. To the best of our knowledge, there is no report on the SOCl$_2$-mediated homocoupling of cysteine in particular and thiols in general.

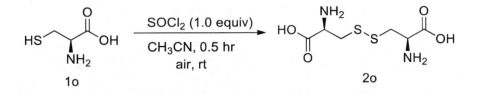

Scheme 1. SOCl₂-mediated synthesis of 3,3'- disulfanediylbis(2-aminopropanoic acid (2o) from cysteine (1o) at ambient conditions.

Herein, we describe a method for the oxidative self-coupling of readily available aliphatic thiols, aromatic thiols and thiol-containing amino acids into symmetrical disulfides using SOCl₂ as a sole reagent. Reactions proceed well in CH₃CN as well as in the solid state at room temperature and pure products in several cases were readily obtained without any chromatographic separations.

RESULTS AND DISCUSSIONS

In order to optimize the reaction conditions 4-methylthiophenol (1a) was used as a model substrate and thionyl chloride as an oxidizing agent. The results of optimization reactions performed under a variety of conditions are summarized in Table 1. Investigations of the model reaction in various anhydrous solvents including dichloromethane, dichloroethane, tetrahydrofuran, DMF, benzene, acetonitrile and hexane suggested that anhydrous acetonitrile was the best medium for the disulfide formation (Table 1 and Entry 3). The next attempt was to arrive at an optimum stoichiometry of SOCl₂ for the reaction. As shown in Table 1, the higher amounts (1.5 equiv.) of SOCl₂ neither lowered the reaction time nor increased the product (2a) yield (Table 1 and Entry 8) significantly. Nevertheless, decreasing the amount of SOCl₂ from 1.0 equiv. to 0.5-0.25 equiv. largely affected the product yield (Table 1 and Entries 9 and 10). Thus, the best result for the reaction is 1.0 equiv. of SOCl₂, 1.0 equiv. substrate in acetonitrile under open air atmosphere.

Table 1. Optimization of reaction conditions for thionyl chloride-mediated oxidation of thiols into disulfides

Entry	SOCl$_2$ (mmol)	Solvent	Time (h)	Yield (%)
1	1.0	Dichloromethane(DCM)	1	75
2	1.0	Dichloroethane(DCE)	0.5	75
3	1.0	Acetonitrile	0.5	>99
4	1.0	Hexane	0.5	90
5	1.0	Tetrahydrofuran	overnight	65
6	1.0	DMF	0.5	75
7	1.0	Benzene	0.5	97
8	1.5	Acetonitrile	0.5	95
9	0.5	Acetonitrile	1.0	50
10	0.25	Acetonitrile	overnight	<50
11	0	Acetonitrile	overnight	NR

[a]All the reactions were performed with 1.0 mmol of 1a and indicated amount of SOCl$_2$ at room temperature under open air atmosphere.

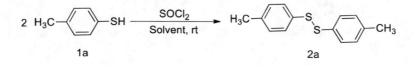

With the optimized conditions in hand (Table 1), the oxidation of various thiols was examined to explore the scope of the reaction. As can be seen from Table 2, different classes of thiols such as aryl thiols (**1a-1j**), alkyl thiols (**1k-1n**) and thiol-containing amino acids (**1o-1q**) were readily converted to the corresponding disulfides in good to excellent yields (Table 2). Moreover, sterically hindered thiols (i.e., **1f-1j**) underwent reaction smoothly and afforded product in good yield (Table 2). This is an attractive advantage of present methodology as several reported protocols are not applicable to oxidation of sterically hindered thiolsubstrates [13-15, 17-21].

$$R^{-S^{-}H} \xrightarrow[\text{CH}_3\text{CN, rt, 0.5 hr}]{\text{SOCl}_2 \text{ (1.0 equiv)}} R^{-S^{-}S^{-}R}$$

Table 2. Substrate scope for the synthesis of disulfides

Entry	Substrate	Product	Time (h)	Yield (%)[b]
1	1a	2a	0.5	99
2	1b	2b	0.5	98
3	1c	2c	0.5	99
4	1d	2d	2	97
5	1e	2e	0.5	99
6	1f	2f	0.5	98
7	1g	2g	0.5	60[c]
8	1h	2h	0.5	97
9	1i	2i	0.5	>99

Table 2. (Continued)

Entry	Substrate	Product	Time (h)	Yield (%)[b]
10	1j	2j	0.5	>99
11	1k	2k	0.5	97
12	1l	2l	0.5	80
13	1m	2m	24	50[d]
14	1n	2n	1	80
15	1o	2o	0.5	99
16	1p	2p	0.5	98
17	1q	2q	0.5	96

[a]All the reactions were performed with 1.0 mmol of thiol and 1.0 mmol of SOCl2 at room temperature.
[b]Isolated yields. c2 equiv. of SOCl2 was used. d2 equiv. of SOCl2 and DCM were used.

In recent time, solvent-free synthesis is gaining importance as a tool for the synthesis of a wide variety of organic compounds[35]. In fact, in a number of cases, solvent-free reactions occur more efficiently than the reactions carried out in solution [36]. Moreover, solvent-free reactions are simple to handle, green and are especially important in industries. Thus, we tried SOCl₂-assisted oxidation of thiosalicyclic acid (**1g**) and cysteine (**1o**) under solvent-free condition at room temperature. To our satisfaction, both reactions underwent readily to give expected products in good yield, *cf.* Scheme 2. Moreover, the reaction rate in the solid state was significantly higher than the solution reactions under the same experimental conditions. Thus, present protocol may be utilized to synthesize S-S bonded proteins and polymers in the solvent-free conditions.

A working mechanism for the preparation of disulfides from thiols is outlined in Figure 1 on the basis of previous literature reports [21, 37]. A nucleophilic reaction between a thiol and SOCl₂ provides a sulfochloridothioite intermediate (**I**), which in turn react with another unit of thiol to form a trisulfide-2-oxide (**II**). As trisulfide-2-oxides (**II**) are highly unstable and known to release SO gas at room temperature, [38] the expected disulfides formed by the decomposition of trisulfide-2-oxide, cf. Figure 1.

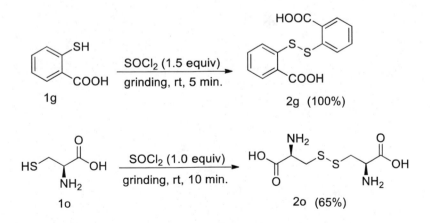

Scheme 2. SOCl2-mediated oxidation of thiols into disulfides under solvent-free conditions.

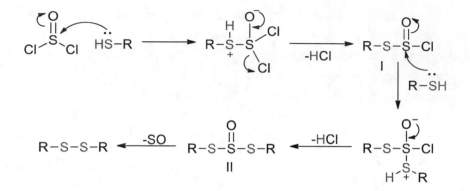

Figure 1. A proposed mechanism for the SOCl2-assisted conversion of thiols into disulfides at ambient conditions.

In summary, a novel protocol is reported for the oxidation of thiols into the corresponding disulfides using $SOCl_2$ in acetonitrile or $SOCl_2$ alone at ambient conditions. Importantly, no over oxidized products were observed. Moreover, products in good to excellent yields are readily obtained by a simple aqueous work-up procedure. We believe that the present protocol may find bright future in the synthesis of S-S bonded proteins and polymers.

EXPERIMENTAL SECTION

General Procedure for the Synthesis of Symmetrical Disulfides

To a solution of thiol (1.0 mmol) in 5 mL of acetonitrile was added thionyl chloride (1.0 mmol) and the resultant mixture was stirred for 30 min. at room temperature. The progress of the reaction was monitored by thin layer chromatography. After completion of the reaction, the reaction solvent was evaporated. Further, the crude product was extracted with water (20 mL) and DCM (3 x 20 mL). The combined organic layers were washed with brine (2 x 15 mL), dried over anhydrous Na_2SO_4, and evaporated in a rotary evaporator under reduced pressure. A reasonably pure product obtained was further purified by recrystallization using hexane–$CHCl_3$ mixture. The purity of the compound was confirmed by melting point and NMR measurements.

General Procedure for the Synthesis of Symmetrical Disulfides by Grinding

Thiol (1.0 mmol) and thionyl chloride (1.0 mmol) were mixed thoroughly using pestle and mortar. The progress of the reaction was monitored by thin layer chromatography. After completion of the reaction the crude product was extracted with water (20 mL) and DCM (3 x 20 mL). The combined organic layers were washed with brine (2 x 15 mL), dried over anhydrous Na_2SO_4, and evaporated in a rotary evaporator under reduced pressure. A reasonably pure product obtained was further purified by recrystallization using hexane–CHCl₃ mixture. The purity of the compound was confirmed by melting point and NMR measurements.

ACKNOWLEDGMENTS

We gratefully acknowledge the financial support of the Department of Science & Technology (DST), India, through INSPIRE Faculty Fellowship [IFA12-CH-62] to P. N.

REFERENCES

[1]   (a) Block, E. *Angew. Chem., Int. Ed. Engl.* 1992, *31*, 1135-1178; (b)Kanda, Y.; Fukuyama, T. *J. Am. Chem. Soc.* 1993, *115*, 8451-8452; (c)Creighton, T. E.; Rog, P. *Prog. Biophys. Mol. Biol.* 1978,*33*, 231-297; (d)Trivedi, M. V.; Laurence, J. S.; Siahaan, T. J.*Curr. Protein Pept. Sci.* 2009, *10*, 614-625; (e) Kadokura, H.; Katzen, F.; Beckwith, J. *Annu. Rev. Biochem.* 2003, *72*, 111-135.

[2]   (a) Cremlyn, R. J. *An Introduction to Organosulfur Chemistry*; Wiley-VCH: New York, 1996; (b) Whitham, G. H. *Organosulfur chemistry*; Oxford University Press: New York, 1995; (c) Kuhle, E.; Klauke, E. *Angew. Chem., Int. Ed. Engl.* 1977, *16*, 735-742; (d) Fletcher, J. M.; Hughes, R. A. *Tetrahedron Lett.* 2004, *45*, 6999-7001; (e) Ulman, A. *Chem. Rev.* 1996, *96*, 1533-1554.

[3]   Reddie, K. G.; Carroll, K.; S. *Curr. Opin. Chem. Biol.* 2008, *12*, 746–754.

[4]   Capozzi, D. C. G.; Modena, G.; Patai, S. *The Chemistry of the Thiol Group*; Wiley-VCH: New York, 1974.

[5]   Joshaghani, M.; Rafiee, E.; Shahbazi, F.; Jafari, H.; Amiri, S.; Omidi, M. *Arkivoc* 2007, 164–172.

[6]   (a) Papadopoulos, E. P.; Jarrar, A.; Issidoides, C. H. *J. Org. Chem.* 1966, *31*, 615-616; (b) Khazaei, A.; Zolfigol, M. A.; Rostami, A. *Synthesis* 2004, *18*, 2959–2961.

[7]   Tajbakhsh, M.; Hosseinzadeh, R.; Shakoori, A. *Tetrahedron Lett.* 2004, *45*, 1889-1893.

[8]   Patel, S.; Mishra, B. K. *Tetrahedron Lett.* 2004, *45*, 1371–1372.

[9]   Iranpoor, N.; Zeynizadeh, B. *Synthesis* 1999, 49–50.

[10]   Raghavan, S.; Rajender, A.; Joseph, S. C.; Rasheed, M. A. *Synth. Commun.* 2001, *31*, 1477-1480.

[11]   Montazerozohori, M.; Karami, B.; Azizi, M. *Arkivoc* 2007, 99-104.

[12]   Shaabani, A.; Ali, M. B.; Rezayan, H. *Catal. Commun.* 2009, *10*, 1074-1078.

[13]   Shah, S. T. A.; Khan, K. M.; Fecker, M.; Voelter, W. *Tetrahedron Lett.* 2003, *44*, 6789-6791.

[14]   Khodaei, M. M.; I. Baltork, I. M.; Nikoofar, K. *Bull. Korean Chem. Soc.* 2003, *24*, 885-886.

[15]   Chauhan, S. M. S.; Kumar, A.; Srinivas, K. A. *Chem. Commun.* 2003, 2348-2349.

[16]   Hajipour, A. R.; Mallakpour, S. E.; Adibi, H. *J.Org. Chem.* 2002, *67*, 8666–8668.

[17]   Zhong, P.; Guo, M. P. *Synth.Commun.* 2001, *31,* 1825-1828.

[18] Demir, A. S.; Igdir, A. C.; Mahasneh, A. S. *Tetrahedron* 1999, *55*, 12399-12404.

[19] Hatano, A.; Makita, S.; Kirihara, M. *Bioorg. Med. Chem. Lett.* 2004, *14*, 2459-2462.

[20] Ali, M. H.; McDermott, M. *Tetrahedron* 2002, *43*, 6271–6273.

[21] Leino, R.; Lonnqvist, J. E. *Tetrahedron Lett.* 2004, *45*, 8489-8491.

[22] Iranpoor, N.; Firouzabadi, H.; Pourali, A. R. *Synlett* 2004, 347-352.

[23] Hayashi, M.; Okunaga, K.; Nishida, S.; Kawamura, K.; Eda, K. *Tetrahedron Lett.* 2010, *51*, 6734-6736.

[24] Lo, W.-S.; Hu, W.-P.; Lo, H.-P.; Chen, C.-Y.; Kao, C.-L.; Vandavasi, J. K.; Wang, J.-J. *Org. Lett.* 2010, 12, 5570-5572.

[25] (a) Lauss, H. D.; Steffens, W. *Ullmann's Encyclopedia of Industrial Chemistry*; Weinheim: Wiley-VCH: New York, 2000; (b) Libermann, D. *Nature* 1947, *160*, 903–904; (c) Relles, H. M. *J. Org. Chem.*1973,*38*, 1570–1574; (d) Zuffanti, S. *J. Chem. Educ.* 1948, *25*, 481.

[26] (a) Pray, A. R.; Heitmiller, R. F.; Strycker, S.; Aftandilian, V. D.; Muniyappan, T.; Choudhury, D.; Tamres, M. *Inorg. Synth.* 1990, *28*, 321-323; (b) Bissinger, W. E.; Kung, F. E. *J. Am. Chem. Soc.*, 1947, *69*, 2158–2163.

[27] Salehi, P.; Zolfigol, M.; Tolami, L. *Phosphorus, Sulfur Silicon Relat. Elem.* 2004, *179*, 1777–1781.

[28] (a) Heravi, M. M.; Derikvand, F.; Oskooie, H. A.; Shoar, R. H.; Tajbakhsh, M. *Synth. Commun.* 2007, *37*, 513–517; (b) Ghammamy, S. Asian *J. Chem.* 2007, *19*917–920; (c) Khodaei, M. M.; Salehi, P.; Goodarzi, M.; Yazdanipour, A. *Synth.Commun.* 2004, 34, 3661–3666; (d) Shirini, F.; Zolfigol, M. A.; Khaleghi, M. *Mendeleev Commun.* 2004, 34–35.

[29] Tajbakhsh, M.; Hosseinzadeh, R.; Shakoori, A. *Tetrahedron Lett.* 2004, *45*, 1889–1893.

[30] Shard, A.; Kumar, R.; Saima, Sharma, N.; Sinha, A. K. *RSC Adv.* 2014, *4* , 33399-33407.

[31] (a) Shaabani, A.; Tavasoli-Rad, F.; Lee, D. G. *Synth. Commun.* 2005, *35*, 571–580;(b) Hajipour, A.; Mallakpour, S.; Adibi, H. *Sulfur Lett.* 2002, *25*, 155–160.

[32] Samanta, S.; Ray, S.; Ghosh, A. B.; Biswas, P. *RSC Adv.* 2016, *6*, 39356–39363.

[33] Choi, J.; Yoon, N. M. *J. Org. Chem.* 1995, *60*, 3266-3267.

[34] Bagi, N.; Kaizer, J.; Speier, G. *RSC Adv.* 2015, *5*, 45983-45986.

[35] (a) Tanaka, K.; Toda, F. *Chem. Rev.* 2000, *100*, 1025–1074; (b) Ohn, N.; Kim, J. G. *ACS Macro Lett.*2018, *7*, 561-565; (c) Uemura, N.; Ishikawa, H.; Tamura, N.; Yoshida, Y.; Mino, T.; Kasashima, Y.; Sakamoto, M. *J. Org. Chem.* 2018, *83*, 2256-2262.

[36] (a) Ye, F.; Haddad, M.; Michelet, V.; Vidal, V. R. *Org. Lett.* 2016, *18*, 5612-5615; (b) Crouillebois, L.; Pantaine, L.; Marrot, J.; Coeffard, V.; Moreau, X.; Greck, C. *J. Org. Chem.* 2015, *80*, 595-601; (c) Konnert, L.; Lamaty, F.; Martinez, J.; Colacino, E. *J. Org. Chem.* 2014, *79*, 4008-4017; (d) Loupy, A.; Petit, A.; Hamelin, J.; Boullet, F. T.; Jacquault, P.; Mathe, D. *Synthesis* 1998, 1213-1234.

[37] Natarajan, P.; Sharma, H.; Kaur, M.; Sharma, P. *Tetrahedron* 2015, *205*, 5578–5582.

[38] Grainger, R. S.; Patel, B.; Kariuki, B. M.; Male, L.; Spencer, N. *J. Am. Chem. Soc. 2011, 133, 5843-5852.*

INDEX

Related Nova Publications

Reactive Oxygen Species (ROS): Mechanisms and Role in Health and Disease

EDITOR: Shannon Wilkerson

SERIES: Chemistry Research and Applications

BOOK DESCRIPTION: This compilation opens with a comprehensive review of heavy metals involves the unifying theme of electron transfer (ET), reactive oxygen species (ROS) and oxidative stress (OS) applied to toxicity, which often arises from pollution.

HARDCOVER ISBN: 978-1-53613-166-6
RETAIL PRICE: $160

Advances in Studies on Enzyme Inhibitors as Drugs. Volume 2: Miscellaneous Drugs

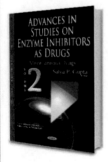

EDITOR: Satya P. Gupta

SERIES: Pharmacology – Research, Safety Testing and Regulation

BOOK DESCRIPTION: The book Advances in Studies on Enzyme Inhibitors as Drugs edited by eminent scientist Satya P. Gupta, covers the most recent development on design and discovery of the most useful drugs acting against several life threatening diseases such as cancer, viral infections and many others in two volumes.

HARDCOVER ISBN: 978-1-53610-505-6
RETAIL PRICE: $230

To see complete list of Nova publications, please visit our website at www.novapublishers.com